Feedback in Analog Circuits

Agustin Ochoa

Feedback in Analog Circuits

Agustin Ochoa
XACT
Vista, CA, USA

ISBN 978-3-319-26250-5 ISBN 978-3-319-26252-9 (eBook)
DOI 10.1007/978-3-319-26252-9

Library of Congress Control Number: 2015955840

Springer Cham Heidelberg New York Dordrecht London

Printed on acid-free paper

Springer International Publishing AG Switzerland is part of Springer Science+Business Media (www.springer.com)

To Dad, Jess, and Basilio, I miss you
To Buck
To Mom
To Sonia, Diane, and David
And always, to Norma

Preface

The path to becoming an analog designer is long, winding, and narrow. It is not well marked and has many exits to other specialties and lines of work. Mostly it is a lonely experience punctuated with periods of intense scrutiny when the design fails to meet a specification. Experience is the key. But where do we get it? We are taught basic tools, such as the Kirchhoff nodal and loop analysis, device models, matrix resolutions, and lots of tricks. What can we simplify, ignore, and approximate? Then we simulate over all imagined conditions and environments. But this process is not satisfying. We use tricks everyone else knows and uses, attributed to someone else, that usually work. Mostly. Matrix analysis produces a jumble of literals that do not easily reveal design trade-offs, and simulation results are even worse with everything done in the background. Developing design insight is difficult.

This book presents a tool to be used in this effort—the Driving Point Impedance/ Signal Flow Graph method (DPI/SFG). This tool, as with all tools, is sharpened with use, and its application becomes secondary allowing the circuit to maintain central focus. With this tool, a circuit is mapped onto a flow graph, a pictorial schematic showing circuit couplings and circuit topology. Loops show feedback, and summers show where and how signals combine. The mapping process with use will not necessitate the intermediate step of drawing the small signal equivalent circuit; and the transfer functions obtained will be in factored form associating pole-zero information with particular nodes and elements–Low Entropy solutions. Such results will aid in the design process and in the development of experience quickly and effectively.

While this book is written for the practicing analog designer struggling to gain insight into a design handed to him from a previous project and asked to fix a particular problem, it is probably of most value to the student, both undergraduate and graduate, before they generate bad habits. It is a tool that once mastered will become a primary approach to their work facilitating the design process as it exposes design dependencies and limitations. It is a book that I wish I had had at an early stage in my career.

How will this book aid in the development of an analog designer? The DPI/SFG process is based on fundamentals. It is grounded in basics and, as such, any conclusion will have a direct path to these fundamentals providing confidence that results are valid. Analysis based on insight lacks this tractability, especially if the insight was not ours. What elements were ignored in the "insight" approximation to your problem? While that simplification may have been appropriate before, is it valid here? And what happens to that approximation when the loop gain goes below unity and we operate in open loop conditions? The DPI/SFG approach provides a basis for examining simplifications systematically to insure that any approximation we make remains valid throughout the range of operation of the design. For example, we use the differential amplifier half-cell equivalents with little concern as to their origin and validity. But what were the assumptions embedded in these equivalents? Primarily symmetry. Does our design have that symmetry? And at what frequency will that high-frequency pole, ignored in this equivalent, become active? We can answer these questions directly using DPI/SFG.

And very quickly along our analog path, we come to feedback which seems to change everything. We soon realize that it is everywhere—an inverter shows the effects of feedback as a kickback transient response when the output switches, causing the input signal to spike, and a single transistor LNA stage has output-to-input capacitive coupling and shares a power line with a power output stage. These examples are parasitic and sometimes may be handled by layout corrections and isolation. Internal device feedback however cannot be corrected and we need to live with it. What about intentional feedback? Now we need to address loading effects and bidirectional signal propagations through circuit elements. Complicating the analysis of feedback is the language, language borrowed from the ideal blocks where life is simple, applied to real circuits where everything mixes. We cannot simply open the loop to find the loop gain or isolate the amplifier to find its open loop gain, A_{ol}. Instead we find that these parameters contain effects of loading, of source impedance, and of reverse loop transmission. This book will show the reader how to handle these properties directly and efficiently.

The path to analog design, while long and winding, has been a most interesting one. My thanks to Don Patterson, Howard Tang, and Roger Gill for walking much of this path with me, to Dr. Jim Early who told me to always sharpen my tools, to Dr. William Shockley who reminded me to count my equations and my unknowns early in problem solving, to Dr. Ruben Kelly who introduced me to DPI, and to Dr. Doug Hamilton, Dr. Gabor Temes, and Bob Baker - excellent teachers and gentlemen.

This work is filled with Pride, Elliot, and Joy(s), Inez, Olivia, Luisa, Amelia, Juliana, and Molly, my grandchildren, and their parents, Sonia and Cliff, Diane and Adrian, and David and Emily.

And a special thank you, with gratitude, to my wife Norma for her constant support and love.

Enjoy!

Vista, CA, USA Agustin Ochoa

Contents

About the Author

Agustin Ochoa received his B.S. from the University of Arizona in 1971, his M.S. from Stanford University (1972) and returned to the University of Arizona for his Ph.D. (1977) in electrical engineering with minors in physics and mathematics. He began his working career at Sandia National Laboratories in Albuquerque, New Mexico, working on CMOS technology development, radiation hardening, and device issues. In 1986 he joined Hughes Aircraft in Carlsbad, California, where he became manager of their BiCMOS development group. His interests throughout his career have been in device physics, in transform mathematics, and in analog circuits. In 1990 he left Hughes Aircraft to join a small analog design startup, Technology Applications Group, leaving his senior device position to become a junior analog designer—the perfect combination of math, device behavior, and analog circuitry, thinking that in a few years he would learn all he needed to be a full analog designer. Then he had to face feedback. After a review of analog texts and many discussions with other designers, he started from the beginning and eventually developed an approach to circuit analysis combining the driving point impedance technique of Ruben Kelly (University of New Mexico) with signal flow graphs (Sam Mason) (DPI/SFG), that allows direct exploration of feedback.

Dr. Ochoa continued developing and applying the DPI/SFG methodology for circuit analysis with particular application to feedback systems in his work at ABB Hafo, Brooktree/Rockwell/Conexant, Fairchild Semiconductor, Xtreme Spectrum, and Zarlink, Ramtron/Cypress Semiconductor and now as an independent consultant at Xtreme Analog Circuit Technology (XACT), working on classic and very low-power analog amplifiers, reference cells, regulators, PLLs, crystal oscillators, active filters, RFID development, and elements of RF.

Dr. Ochoa has taught at the University of New Mexico and at the University of California at San Diego as an adjunct professor and published and lectured on feedback analysis and analog design since 1994. He has been awarded six patents with another six pending. And 24 years later, he is still learning analog design.

Chapter 1
The Concept of Feedback: Why We Care and Complications

Abstract Feedback has long been used in the design of analog systems and circuits to provide system stability from variations in device parameters and in temperature. The understanding, design, and analysis of feedback systems have generally followed insight gained from consideration of ideal systems along with approximations of real systems to fit into the ideal form. This approach is not always successful in gaining understanding as loading and feedforward effects become significant. In this text, we explore an approach that reveals the intrinsic properties of feedback, allowing proper exploration to design with the direct handling of loading and of feedforward effects so that imposing simplifications and approximations early on are not needed. A unified approach is developed to general circuit analysis and to circuits including feedback in particular, in a complete and natural way that aids in developing understanding of design and of feedback systems in particular.

Introduction

Feedback is an important tool in analog circuit design in which a sample of the output signal of an amplifying cell is combined with the input signal and re-amplified as shown in Fig. 1.1. W. Black first described this design process in 1932 [1]. His purpose was to produce a system having reduced sensitivity to both device parameters and to temperature variations. He succeeded remarkably well, but the process for analysis and design as normally presented has remained mysterious and incomplete. The literature of the analysis of feedback circuits and systems is full of approximations and with discussions that are either guided by experience and insight or follow a fully mathematical modeling transformation of circuit variables to matrix algebra using Kirchhoff relations. Neither of these approaches, experience based or full nodal analysis, makes developing a sense for the topic straightforward. The literature is confusing with inspired insights and complex mathematical relations being difficult to interpret. This text presents a unified approach to general circuit and to feedback circuit analysis based on driving point impedance and signal flow graphs (DPI/SFG) [2] and will clarify much of the

A. Ochoa, *Feedback in Analog Circuits*, DOI 10.1007/978-3-319-26252-9_1

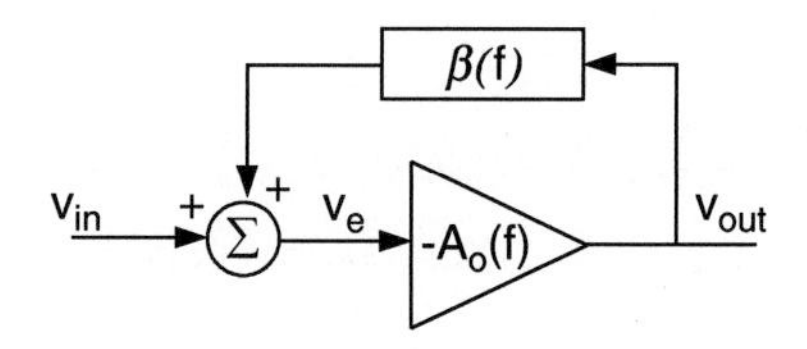

Fig. 1.1 Ideal feedback system

problems encountered in the feedback thesis and aid in the development of insight to analog design.

Our purpose here is to develop and demonstrate a systematic approach to the analysis of circuits, one that is applicable to feedback circuits in particular. The signal flow graph-based approach gives the designer a tool that will enable him to analyze circuits systematically. Using this process, the circuit designer will develop experience with circuit topologies and gain familiarity with the use of mathematical modeling, always keeping the circuit as the focus of his arguments.

Using DPI/SFG, we solve for fully loaded transfer and loop gain functions of single and multiple loop systems. A benefit of this approach is that results appear in factored form—properties of the responses can then be associated to particular components of the circuit providing guidance for design. In this form, further analysis using math engines such as MatLab or Xcel to explore effects of design variables is facilitated. Interpretation of flow graph topologies guides the development of SPICE-type simulation setups to produce fully modeled feedback designs. A series of analog systems are then discussed using the DPI/SFG approach as the basis for analysis: Amplifier compensation, crystal oscillator, phase-locked loop systems, bandgap reference voltage design, G_{m}–C filters, and system noise analysis are demonstrated.

This chapter presents the concept of feedback, why we use feedback in analog circuits, and identifies the basic problems encountered in the analysis and design of such systems. The driving point impedance/signal flow graph (DPI/SFG) procedure is fully developed in Chap. 2, starting from fundamental concepts of circuits and small signal modeling. Succeeding chapters use DPI/SFG to find and explore a number of circuit properties of interest to the analog designer: transfer functions, port impedances, noise response, and modeling. The concept of loop gain is explored in the feedback chapter, Chap. 5—the ambiguities of this concept are explained as arising from the fact that system loop gain function comes from analysis and is itself not a separable property of the nonideal analog circuit. As such it is easy to confuse the loop gain function with functions that fit the feedback form that arise from the analysis. A fundamental definition for finding the proper loop gain function is developed, providing a consistent approach for analysis and for simulation, making the process self-consistent and intuitively appealing. Chapter 6 presents a number of applications starting with simple current mirrors and current and voltage references and then more complex circuits showing loop gain response and compensation. We then present the fundamentals of phase-locked loops and switched mode power supply designs and finish with a discussion of G_m–C filters.

A consequence of analysis using DPI/SFG is that the process consists of working with sub-circuits and mapping the properties of these onto a flow graph as we walk through the design. This process allows the design engineer to literally create the flow graph by inspection of the circuit without the use of intervening small signal equivalent circuits. Very few of these equivalent circuits are used in this text and then, only to aid in the explanation and development of the procedure.

The Problem with Open-Loop Design

The "open-loop" design shows marked sensitivity to circuit elements, power level, and temperature. Variations in device parameters, a resistance or the current gain of the bipolar transistor, directly modify the system response. Such a design practice seriously impacts product yield and limits the use environment. Integrated analog circuits in particular would yield poorly as circuit elements cannot be individually selected to meet a design specification. An example of a simple open-loop amplifier is shown in Fig. 1.2a. The amplifier transfer function using the simplified transistor model in Fig. 1.2b is given by Eq. (1.1). The transistor is represented by a small signal model, here using only its common emitter input impedance h_{ie} and current gain h_{fe}. The voltage gain is a function of these transistor parameters and of circuit resistance values r_s and r_l. Each of these parameters is affected by voltage, temperature, and manufacturing variations, making the voltage gain function vary over a wide range. The base biasing, not shown in Fig. 1.2, will also affect gain as it varies with power supply and temperature, changing the operating point of the transistor. This is not a good design practice for high volume production:

$$v_{\text{out}} = \frac{-v_{\text{s}} \cdot h_{\text{fe}} \cdot r_{\text{l}}}{r_{\text{s}} + h_{\text{ie}}} \tag{1.1}$$

Discrete design techniques "fix" these sensitivity problems by adding a "degeneration" resistor at the emitter as shown in Fig. 1.3a. Using the small signal model in Fig. 1.3b, we generate the transfer function as given in Eq. (1.2):

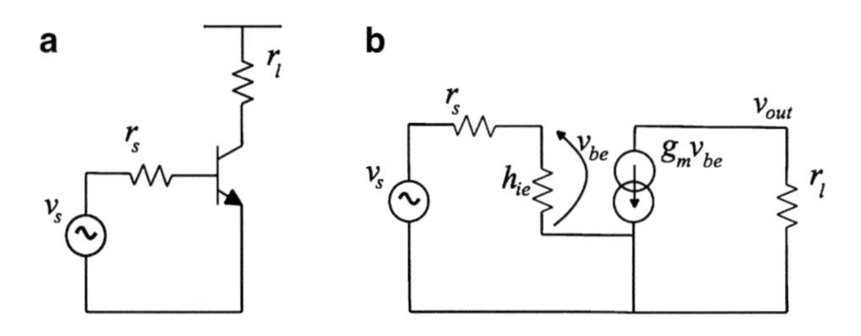

Fig. 1.2 (**a**) Simple open-loop amplifier, (**b**) circuit small signal equivalent model

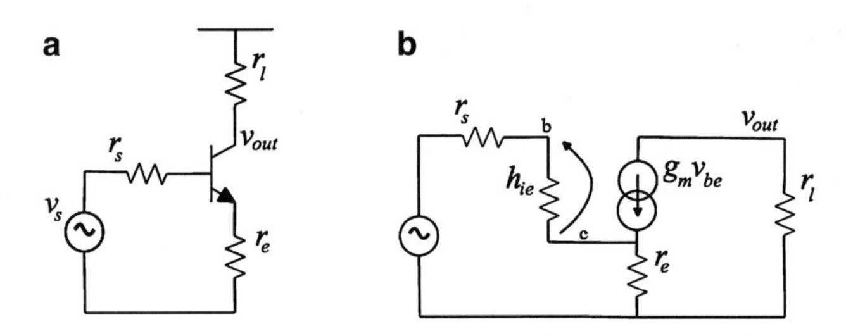

Fig. 1.3 (**a**) Adding degeneration resistance R_e to amplifier, (**b**) circuit small signal equivalent model

$$v_{\text{out}} = \frac{-v_{\text{s}} \cdot h_{\text{fe}} \cdot r_{\text{l}}}{r_{\text{s}} + h_{\text{ie}} + r_{\text{e}} \cdot (h_{\text{fe}} + 1)} \sim \frac{-v_{\text{s}} \cdot r_{\text{l}}}{r_{\text{e}}} \tag{1.2}$$

By making r_e in the denominator of Eq. (1.2) large enough so that r_s and h_{ie} become less significant, the gain relation is made less dependent on the transistor parameters h_{ie} and h_{fe}, and the circuit gain becomes a ratio of resistors r_l to r_e. If the resistors are of the same physical type, they will track in an integrated circuit design with process and temperature, stabilizing the design response against resistor variations as well. This fix is actually a feedback fix. The voltage at the emitter in Fig. 1.3 is proportional to the output signal current in r_l, making the emitter voltage proportional to the output signal. Since the transistor responds to the voltage difference between base and emitter, a fraction of the output voltage is combined with the input signal by transistor action and processed by the amplifier. The result is that the gain function is made less sensitive to the particular parameters of the transistor. Replacing the transistor in a discrete design by another device from another production lot, manufacturer, or even device design may still produce a gain relation within specification.

The Ideal Loop: Black's Feedback Relation

Figure 1.1 shows the block diagram for an ideal feedback system. The amplifier output is scaled and returned to the input where it is *added* to the input signal. The sum is then reprocessed by the amplifier. The summer is usually shown as a differencing block, creating a phase reference offset of 180° in the description of the loop behavior external to the circuit blocks. Here we avoid this offset (use a summer, not a subtractor) and account for all phase shifts directly in the blocks themselves. The amplifier input in this case will have an inverting input, placing the 180° phase shift in the amplifier and feedback net where it occurs.

Block diagram algebra used in analyzing Black's feedback system in Fig. 1.1 is very direct: Signals proceed in the direction of the arrows and are multiplied by the block contents. The ideal summer produces the signed sum of its inputs. From these simple rules, we can write for v_e, the error signal at the output of the summer, and for v_{out} at the output of the amplifier, the relations in Eq. (1.3):

$$\begin{aligned} v_{\mathrm{e}} &= v_{\mathrm{in}} + \beta \cdot v_{\mathrm{out}} \\ v_{\mathrm{out}} &= -A_{\mathrm{ol}} \cdot v_{\mathrm{e}} \end{aligned} \tag{1.3}$$

Solving (1.3) for v_{out} as a function of v_{in}, we get Black's feedback system transfer function as given in Eq. (1.4):

$$v_{\mathrm{out}} = \frac{-A_{\mathrm{ol}}}{1 - (-\beta \cdot A_{\mathrm{ol}})} \tag{1.4}$$

$-A_{ol}$ is the amplifier's "open-loop" gain, and β is the feedback factor specifying the amount of output that gets returned to the input. The product of the loop blocks β and $-A_{ol}$ is the system "loop gain" LG.

For large loop gains, the output/input transfer function goes to $\sim -1/\beta$. The system gain is seen to become independent of the amplifier parameters and defined entirely by the feedback factor β. As was shown in the emitter degeneration example in Fig. 1.3, this factor can be made a ratio of passive components of like kind that in integrated circuits will match with processing and will track with temperature, thereby stabilizing the transfer function. This improved behavior is seen to come at the expense of amplifier gain.

Analyzing ideal systems composed of unilateral blocks with no frequency or loading effects is straightforward and simple. The analysis of a *real* system will deviate significantly from this modeling due to loading, reverse transmission of signals through the amplifier and the feedback element, and variations of circuit element properties with frequency. System properties such as open-loop gain, feedback factor, and nodal impedances become ill-defined with components of both amplifier and feedback net contributing to their effective responses. These properties complicate analysis and must be fully taken into account to find the proper functions so simply modeled above.

Complications in Feedback Design

While feedback can be used to stabilize a design to parameter and temperature variations, it can also produce unwanted behavior. The feedback transfer function in Eq. (1.4) shows the basic problem. The assumption in this relation has been that the signal "through the loop" remains negative—that the phase shift is 180°. Neither gain nor phase is constant over the whole frequency spectrum. Should the phase in the loop increase by an additional 180° the feedback will become "positive," the returned signal will be in phase alignment with that of the input signal, causing it to increase in magnitude, and the system may become unstable breaking into oscillations or clamping at one of the power rails. The design of feedback systems is not always straightforward, using new concepts "loop gain" and "phase margin" which require much more than knowledge at DC.

The feedback situation leading to potential system instability is aggravated in circuits that need to operate at high frequencies. MOS devices have internal parasitic capacitive couplings between output and input nodes, c_{gd}, creating an unintentional parasitic feedback loop (bipolar transistors and vacuum tubes will have similar capacitive coupling paths between input and output ports). Low-frequency designs can usually safely ignore these couplings, but in high frequencies, the capacitive admittance increases, and sufficient energy can then couple from output to input to produce a positive feedback situation.

And the story continues. Other feedback paths exist that are more difficult to find and analyze. Power is supplied to an amplifier block through supply lines having distributed impedances, feed input and output stages, as well as biasing circuitry. Current delivered to the load through the output stage is supplied through these lines, putting a signal on the power lines that can couple back to the input stage amplifier. Sharing power supplies between cells couples signals as does sharing substrate material and signal lines crossing or coming close to each other. These latter unwanted couplings are managed through careful layout, board, and power distribution design.

Real Loops

From this simple discussion of idealized feedback systems, we identified some basic feedback variables—loop gain, error voltage, feedback factor—and obtained the system transfer function. In this section, we add properties of real systems to the circuit blocks to identify issues in the design of real feedback systems. An electronic feedback system having a response approximating the transfer function for the idealized system in Fig. 1.1 is shown in Fig. 1.4. Analysis for this system begins by assuming ideal properties for the amplifier—that it have infinite input impedance and zero output impedance. The error voltage v_e, seen to be the differential voltage at the input to the amplifier, is given in Eq. (1.5):

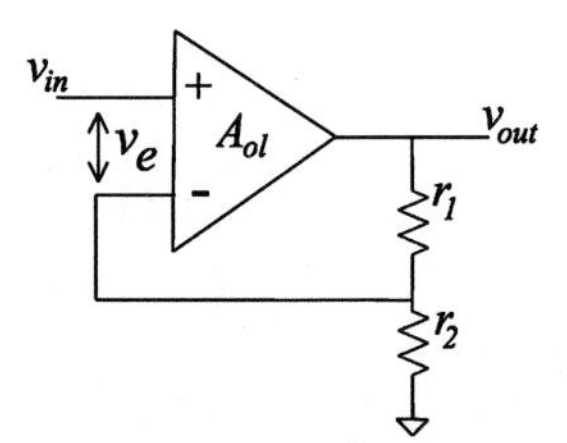

Fig. 1.4 Operational amplifier feedback system

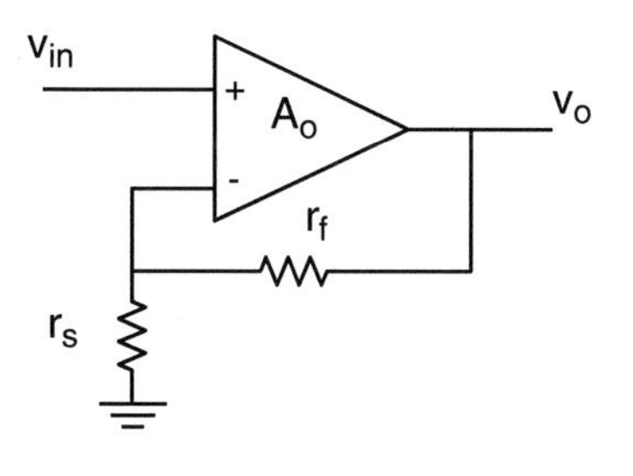

$$v_e = \frac{v_{in}}{1 + \frac{r_2}{r_1 + r_2} \cdot A_{ol}} \tag{1.5}$$

A large loop gain (A_{ol}) will drive this error voltage to ~0, making the voltage at the amplifier inverting input node approximately equal to the voltage at the non-inverting input, v_{in}. Since the current into the amplifier block is zero due to the assumed infinite input impedance, we can write directly for the output voltage as in Eq. (1.6):

$$v_{out} = v_{in} \frac{r_1 + r_2}{r_2} \tag{1.6}$$

We see the feedback signature in this final result—a removal of the particulars of the active element (amplifier).

The next order of analysis recognizes the effects of loading, frequency, and bilateral signal transmission on the system response. Transfer function parameters A_{ol} and β and therefore loop gain LG, are complex functions of frequency. As such the "feedback" may change from negative to positive for some frequency range, changing the character of the response dramatically. Using a simple 2-port model for the amplifier in Fig 1.4, (y_i, g_o, G_m), we generate a system flow graph, Fig. 1.5, where $g_1 = 1/r_1$, $y_{2i} = 1/(y_i + y_2)$, $g_2 = 1/r_2$, $y_{ld} = 1/(r_1 + 1/y_{2i})$, and $A_{ol} = G_m/g_o$.

Frequency behavior is contained in amplifier model parameters y_i and G_m and in the load y_l. For this simple example, no additional load is assumed. Only the feedback network loads the amplifier. The transfer function is found by solving the flow graph in Fig. 1.5 for v_o, Eq. (1.7):

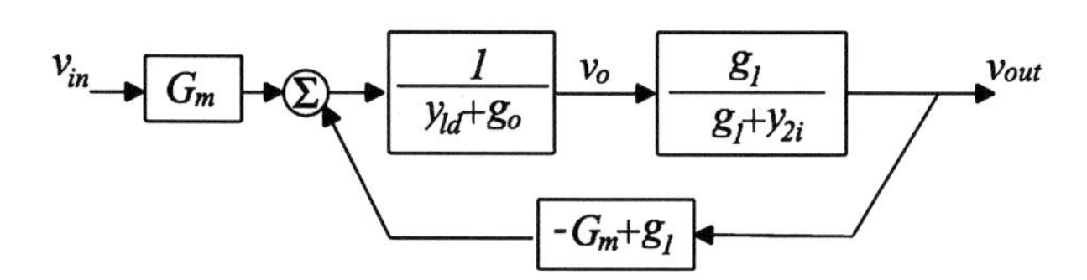

Fig. 1.5 System flow graph for feedback amplifier in Fig. 1.4

$$v_o = v_{in} \cdot G_m \frac{\frac{1}{g_o + y_{ld}}}{1 - \frac{1}{g_o + y_{ld}} \frac{g_1}{g_1 + y_{2i}} (-G_m + g_1)} \tag{1.7}$$

At low frequencies where the loaded amplifier gain is high, we get a gain given in Eq. (1.8):

$$h(f) \sim \frac{(g_1 + y_{2i})}{g_1} \sim \frac{r_1 + r_2}{r_2} \tag{1.8}$$

From Eq. (1.7) we see that the full transfer function has a singularity at that frequency where the denominator goes to zero. Since a real amplifier gain will "roll off" at high frequencies, the function subtracted from "1" in the denominator will also roll off and cross unity gain at some frequency, f_{unity}. For frequencies near f_{unity}, the phase relation will define how closely the denominator approaches zero, where the transfer function becomes unbounded. For frequencies near this condition, magnitude 1 and phase near 0°, the magnitude of the denominator becomes small, and the transfer function increases—the system becomes sensitive to changes in the steady-state input signal, generating responses characterized by ringing transients or, worse, by sustained oscillations. From this observation, we establish that the denominator function defines system stability and that a measure of this stability is how closely the denominator approaches zero in the system transfer function. *Phase margin is defined as the angle that the loop gain differs from zero, where the loop gain magnitude becomes one at* f_{unity}.

It is important to note the distinction between this definition for phase margin and that generally found in the literature where phase margin is referenced to 180° rather than 0°. This is a consequence of "adding" the feedback signal to the input signal instead of "subtracting" as is normally done. The use of a summer puts all phase shift relations onto the electronic subblocks where they occur, providing a consistency in definition. This definition for phase margin is equivalent to that obtained using the subtraction since phase margin is a difference angle.

While we see aspects for controlling the design—keeping the denominator from getting too close to zero—we do not at this point have the system loop gain. So what is loop gain? We define this soon.

In the simple example presented above, we can see that the approach handled two common problematic issues without difficulty: loading and bilateral properties of elements. The open-loop characterization of the amplifier gain, A_{ol}, is appropriate only if the feedback element and any circuitry attached to the cell output do not affect this characterization significantly. Loading effects are minimized with low-output impedance amplifiers. Bilateral elements allow energy to pass in both directions so that feedforward signals through the passive elements may modify the response of the cell. The amplifier itself may also be bilateral.

The analysis and design of real circuits must account for these three issues: frequency effects, loading, and bilateral signal transfer.

The Driving Point Impedance/Signal Flow Graph Approach

The design of ideal feedback systems is well defined. *We must maintain negative feedback for loop gain magnitudes around unity*. If that were the whole story, then this book would not be needed. The problem lies in identifying what is meant by feedback before we can begin to meet the loop gain requirements. The design of real feedback systems is less well defined. For the ideal blocks in Fig. 1.1, it is correct to open the feedback loop in determining loop gain as the function in the denominator of the transfer function. Here the blocks are unidirectional and do not load the amplifier. An intuitive definition for feedback is readily attained. This is not generally true for real circuits such as that in Fig. 1.4 where the action of opening the circuit loop to obtain factors in transfer functions such as those in Eqs. (1.1) and (1.2) change loading and transmission paths, yielding incorrect results. Opening the loop in some circuits such as that in Fig. 1.3 makes the circuit nonfunctional, and no valid analysis can be obtained. Yet there is an implied ability to "open" loops in the term used, A_{ol} for open-loop gain. What we need to "open," or more accurately "zero," is the feedback signal and not necessarily the "loop."

It is easy to misinterpret a flow graph loop such as that in Fig. 1.5 with the loop gain function. The flow graph algebra results in a transfer relation having the feedback "form." To properly generate the loop gain function, we must design the graph with loop gain information included from the start [3]. To do this, we separate signals in the feedback system in Fig. 1.4 as shown in Fig. 1.6.

The currents i_1 and i_2 are sourced by independent voltage sources v_1 and v_2. DC conditions are held on the capacitors and isolation at all but very low frequencies imposed by the inductor, all large valued. These currents can each be expressed as the sum of two components: i_{xx}, current from source "x" to ac ground, a "self" current, and i_{yx}, current from source "x" into source "y," a "cross" current. With this division, we generate the system flow graph shown in Fig. 1.7.

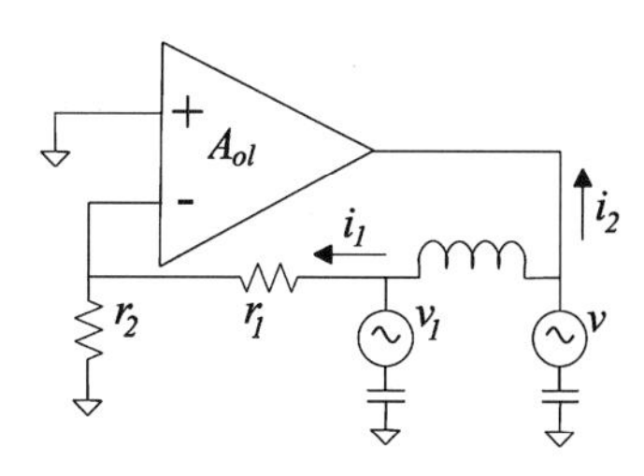

Fig. 1.6 Schematic for generating terms in loop gain function

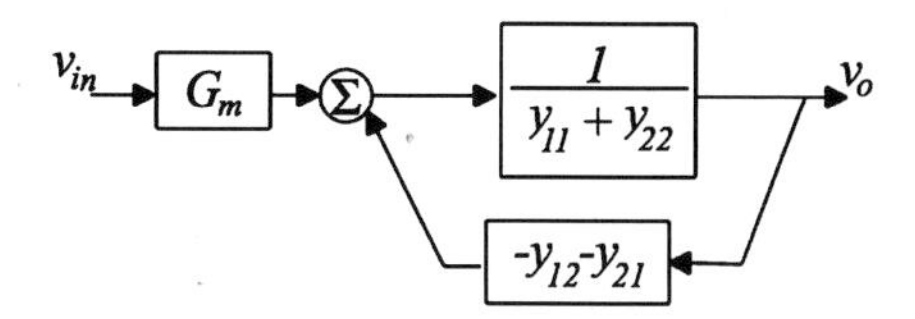

Fig. 1.7 Small signal model for split node in Fig. 1.6

With this construction, we define the system loop gain function properly as in Eq. (1.9):

$$Loop\ Gain = \frac{-(y_{21} + y_{12})}{y_{11} + y_{22}} \tag{1.9}$$

Loop gain becomes the negative of the sum of the cross-conductances divided by the sum of the self-conductances.

Equation (1.9) and this definition of loop gain are fully developed in Chap. 6.

Summary

Classic analog design texts usually approach feedback with examples of feedback circuits examined sequentially and show how they "fit" into the basic relations. The examples either guide the student through an analysis using experience not possessed by the student or the beginning design engineer or more formally follow a full mathematical modeling two-port/matrix approach where the complexity quickly overpowers the novice and experienced designer with complex algebraic expressions. Neither of these approaches efficiently develops insight in the novice designer. In the experience-based analysis, the student is presented with a set of approximations or tricks that he may later recall for similar topologies while perhaps not fully understanding the reasoning behind them. The full nodal analysis presents the process for systematic computer analysis, generating results for all but the simplest designs too complex to provide insight. The process does work as we have a large number of successful analog designers busy producing working designs, but it does not work all that well as many designers have trouble defending their designs and quickly resort to simulation-based incremental improvements. The novice and senior designer alike is left with many unanswered questions.

This book addresses these problems in the analysis and design of real electronic feedback systems. A systematic approach to analysis based on the principles of equivalent circuits, Thevenin and Norton transformations, and signal flow graphs is given. This process reduces the need for direct experience with a circuit core for insight-guided analysis. A process for mapping a circuit onto a flow graph will be developed. Once the flow graph is created, the analysis becomes to an extent mechanical in the use of flow graph algebra and flow graph constructions. Transfer function relations are separable from circuit topologies. Open-loop quantities are obtained from the graph, the mathematical domain that is ideal. This process fully accounts for the elements that make classic analysis difficult—loading and bilateral transmission effects.

Flow graphs provide additional analysis advantages. Once the flow graph is created in a primitive form, voltages at every node are easily obtained. And the graph is expandable to provide branch currents and other circuit properties such as power and port impedances. We shall see how to include changes to the basic circuit topology easily, not requiring one to start the analysis completely over again. New loads or feedback elements are quickly included in the analysis. With another slight modification to the graph, we can obtain port impedances. Noise signals can be translated from node to node, making noise analysis direct and transparent. The process is applied independent of technology so that bipolar, MOS, and vacuum tube circuits (add your favorite technology) show similarities and strong unity. And finally, flow graphs can be used to guide simulations to provide insight into circuit behavior by generating transfer function factors in a decomposition analysis. The design process can be broken up into subblocks by identifying individual poles or zeros dependent upon unique circuit elements—"low entropy" results.

While this text presents circuit analysis using flow graphs, formal graph analysis is not the focus—instead an intuitive approach is developed for graph algebra manipulations. This is not a text on flow graphs but a text where this tool is used to increase the awareness of the circuit topologies and their key features. A direct mapping of circuit to flow graph is developed. Once the graph is identified, analysis proceeds using simple graph manipulations and reductions to obtain the desired transfer function in the Laplace domain. We flash forward to describe a complete feedback analysis where loading and feedforward effects are included in the analysis directly. The key feedback relation, loop gain given in Eq. (1.8), is seen to be a symmetrical relation involving two loops. The set up in Fig 1.6 will be fully developed and expanded to include multiple and nested loops and differential designs as a proper and complete method for obtaining loop gain.

References

1. H.S. Black, *Stabilized feedback amplifiers* (Bell. Syst. Tech. J., 13, 1934), pp. 1–18
2. A. Ochoa, Properly simulating the open loop. in *Mid West Symposium on Circuits and Systems* (1998)
3. A. Ochoa, Loop gain in feedback circuits: a unified theory using driving point impedance, in *Mid West Symposium on Circuits and Systems* (2013)

Chapter 2
Driving Point Impedance and Signal Flow Graph Basics: A Systematic Approach to Circuit Analysis

Abstract The driving point impedance/signal flow graph (DPI/SFG) methodology is based on fundamental analysis techniques—equivalent circuits and superposition, which along with a direct mapping of circuit relations onto a signal flow graph results in a direct and systematic analysis process. Following this approach leads to factored solutions for direct manipulation of the design to produce a desired shaping of the design response. The DPI approach divides the analysis problem into two simpler subproblems, one for finding the current into an AC ground and the other for finding the port impedance. A nodal voltage response is then given as the product of these two results, $v_{out} = i_{sc} \times \mathrm{DPI}_{out}$. Signal flow graph analysis is a technique for representing system algebra pictorially. The combination of these tools provides a unified approach to general circuit analysis. This chapter develops the DPI/SFG analysis methodology and demonstrates its application using a variety of circuits to show its versatility and power.

Driving Point Impedance

The driving point impedance technique to circuit analysis [1] first solves two simpler sub-circuits and then combines the results to produce the full solution to the original circuit. For one sub-circuit, we place a short to ac ground at the desired output node "x" and solve for the current into this short, the short-circuit current, I_{scx}. The second sub-circuit is designed to find the driving point impedance into this node, DPI_{outx}. The voltage response at this node is the product of these two components, $V_{outx} = I_{scx} \times \mathrm{DPI}_{outx}$. Consider an arbitrary single-input-generalized circuit as represented in Fig. 2.1a. For this discussion, we present the problem as a DC excitation to a passive net for simplicity. We attach a variable auxiliary power supply to the node "x," V_x, and monitor the current that the net sources *into* the auxiliary supply, I_x. Since the circuit is linear, the current sourced to the auxiliary voltage source is also linear with the auxiliary voltage source V_x, Fig. 2.1b. We can dial the applied voltage V_x up and down, watching the output current until the output current becomes zero. At this value of V_x, the net is not loaded by the auxiliary source, and removing the source will not perturb the net. The output node

A. Ochoa, *Feedback in Analog Circuits*, DOI 10.1007/978-3-319-26252-9_2

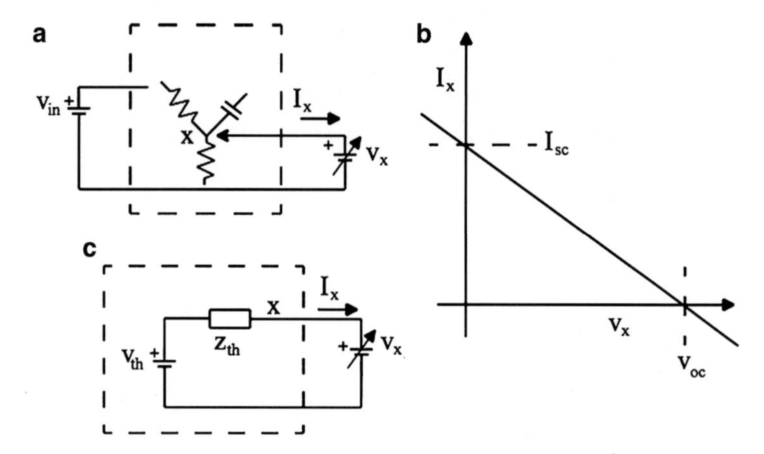

Fig. 2.1 (**a**) Arbitrary circuit with auxiliary source V_x at output, (**b**) linear current response at output nodes as a function of V_x, and (**c**) Thevenin equivalent at "x" for circuit with applied auxiliary source

remains at this value of V_x with I_x equal to zero—the open-circuit condition, V_{oc}. The problem then is to find this value for the auxiliary source that results in zero output current.

To find this balanced condition, we first replace the circuit with its Thevenin equivalent circuit, Fig. 2.1c, and solve for the zero output current condition to obtain Eq. (2.1). Here we recognize that the quotient of the Thevenin voltage to the Thevenin impedance, V_{th}/Z_{th}, is the current that would flow into a short circuit at the output, I_{scx}, and that the Thevenin impedance is the driving point impedance at the port, DPI_x.

$$I_{\mathrm{x}} = \frac{V_{\mathrm{th}} - V_{\mathrm{x}}}{Z_{\mathrm{th}}} \\ \Rightarrow V_{\mathrm{x}} = I_{\mathrm{scx}} \cdot \mathrm{DPI}_{\mathrm{x}} \tag{2.1}$$

The method of driving point impedance/signal flow graph utilizes Eq. (2.1) repeatedly as will be seen later. An alternative argument to Eq. (2.1) utilizes a Norton equivalent to the port "x" and solves for I_{scx} and DPI_x. We will use either of these circuit constructs to generate the DPI quantities, selecting the source based on how it may facilitate the analysis.

While this argument was presented for a passive circuit with DC excitation, the same argument holds for AC excitation and for linear circuits with active elements. The "short-circuit current" for the AC case is into an AC ground. The significant result is that *the output voltage is given as the product of the short-circuit current and the driving point impedance* at the output port. *ac* signals will be represented in lower case to differentiate from the *DC* signals. A particular advantage of this approach is that in the process of solving a circuit response, with proper selection of the node "*x*," we may disable the feedback and analyze sub-circuits without feedback. We capitalize on this property to solve feedback circuits with little extra effort than that for solving non-feedback circuits.

Some Simple Applications of the DPI Law

Let us apply the DPI method to a few simple problems to demonstrate the process. The circuits we begin with are meant more as examples of known results to focus on the process and not on the circuits. For the methodology to have merit, we must obtain advantages over other methods of analysis. The DPI process has the advantages of (1) being systematic in application. New circuits will be approached in the same way so that a priori knowledge or insight is not required. (2) Results are in factored form that aid in the design function. The effect of particular nodal impedance or of a current division on the resulting transfer relation will appear as a factor in the analysis so that individual pole and zero locations can be identified and modified. (3) The analysis is simplified compared to that of writing a complete set of nodal relations and solving by substitution or matrix methods. And (4) this approach makes analyzing feedback circuits direct and unambiguous.

First consider the resistor voltage division problem in Fig. 2.2a. For the short-circuit current, we place a short to ground at V_{out} as shown in Fig. 2.2b. This construct removes r_2 from consideration, simplifying the problem to finding the current through a single resistor r_1 as given in the top equation in the Eq. (2.2) set.

For the DPI response at the output node, we modify the circuit to that in Fig. 2.2c with the input voltage source v_s set to its internal impedance, 0 V and 0 Ω. The port driving point impedance for this case is seen to be that of two resistors in parallel. Parallel admittances are easier to handle, adding the values directly and taking the reciprocal of this sum for the impedance as in Eq. (2.2) second equation rather than forming the ratio of the product with that of the sum required by remaining in

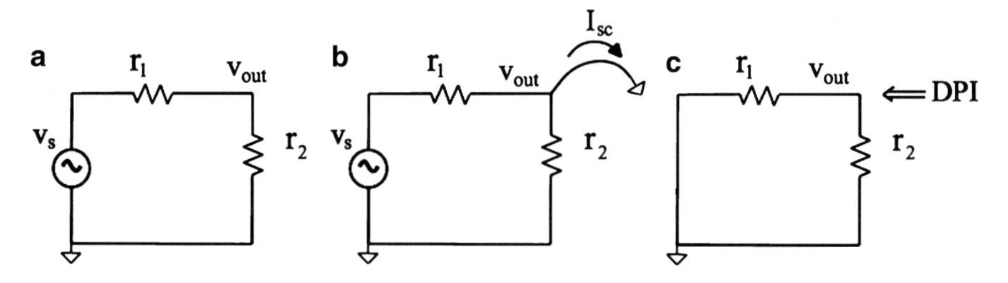

Fig. 2.2 (**a**) Voltage division circuit for demonstration of the DPI approach, (**b**) construction for I_{sc}, and (**c**) construction for DPI

impedance terms. The output voltage is found as the product of these factors, bottom equation in Eq. (2.2).

$$\begin{aligned} I_{sc} &= V_s \cdot g_1 \\ \text{DPI} &= \frac{1}{g_1 + g_2} \\ V_{out} &= V_s \cdot g_1 \cdot \frac{1}{g_1 + g_2} = V_s \cdot \frac{r_2}{r_1 + r_2} \end{aligned} \tag{2.2}$$

The use of conductance simplified the algebra. In general, the admittance or the impedance form is used throughout the text where one representation simplifies the analysis over the other. The final line shows the normal form for this result.

While this example is trivial, it serves to demonstrate the process for finding a node voltage using the DPI technique: (1) Find the current into a short placed at the output node; (2) find the driving point impedance at the node first setting all independent sources to their internal impedance, with voltages becoming shorts and currents becoming opens; and (3) take the product of these two factors. We analyze two simpler circuits than the original and combine results.

For a more interesting circuit, we look at the single transistor amplifier shown in Fig. 2.3a. We first find the circuit's voltage transfer function using the usual Kerckhoff nodal analysis and then again using the DPI method. This amplifier will be revisited in the next section using the full DPI/SFG approach. Small signal analysis generally proceeds by replacing the active element with its small signal equivalent model as shown in Fig. 2.3b where here we use only the input impedance and the transconductance elements to model the transistor. We then write a set of nodal and loop relations and solve by substitution or matrix algebra. Summing the currents into base and collector nodes and using conductance wherever we get simpler expressions, we write relations in Eq. (2.3). Note that $y_{ie} = 1/h_{ie}$ and $g_m = h_{fe}/h_{ie}$.

$$\begin{aligned} (v_s - v_b) \cdot g_s &= (v_b - v_{out}) \cdot g_f + v_b \cdot y_{ie} \\ (v_b - v_{out}) \cdot g_f &= v_{out} \cdot g_l + v_b \cdot g_m \end{aligned} \tag{2.3}$$

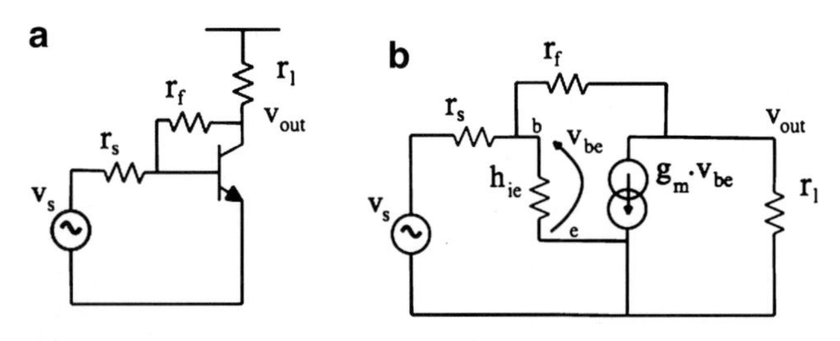

Fig. 2.3 (**a**) Single transistor feedback amplifier and (**b**) equivalent circuit using a simple small signal model for the transistor

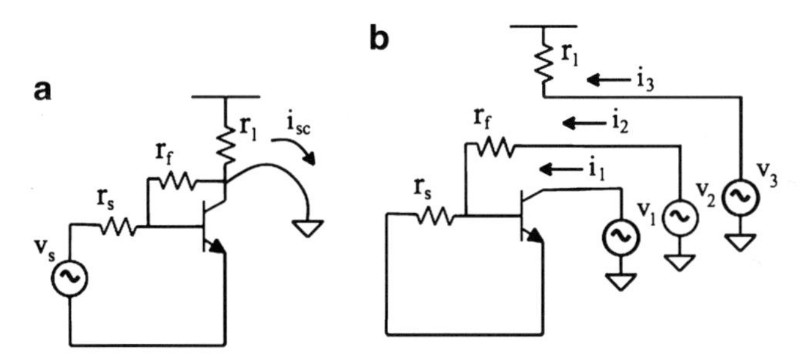

Fig. 2.4 Modifications to the circuit in Fig. 2.3a for finding (**a**) short-circuit current and (**b**) driving point impedance at the output node

This set of equations in two unknowns can be solved by substitution to produce Eq. (2.4).

$$\frac{v_{\text{out}}}{v_{\text{s}}} = \frac{(g_{\text{f}} - g_{\text{m}}) \cdot g_{\text{s}}}{(y_{\text{ie}} + g_{\text{f}} + g_{\text{s}}) \cdot (g_{\text{f}} + g_{\text{l}}) - g_{\text{f}} \cdot (g_{\text{f}} - g_{\text{m}})} \tag{2.4}$$

It is clear that the result in Eq. (2.4) is a jumble of literals providing little immediate design insight. The limiting result with a very large transistor transconductance, g_{m}, becomes Eq. (2.5).

$$\frac{v_{\text{out}}}{v_{\text{s}}} \sim \frac{-g_{\text{s}}}{g_{\text{f}}} = \frac{-r_{\text{f}}}{r_{\text{s}}} \tag{2.5}$$

To solve this circuit response using DPI, we first apply an ac short at the collector in Fig. 2.4a and then solve for the current out of the node and into the ac short to ground, i_{sc}:

$$i_{\text{sc}} = \frac{v_{\text{s}}}{r_{\text{s}} + \dfrac{1}{g_{\text{f}} + y_{\text{ie}}}} \cdot \left(\frac{g_{\text{f}}}{g_{\text{f}} + y_{\text{ie}}} + \frac{-g_{\text{m}}}{g_{\text{f}} + y_{\text{ie}}} \right) \tag{2.6}$$

The construct of adding the ac short at the collector removes the circuit feedback allowing us to walk through the circuit from input to output adding a series of factors to the evolving solution. The first factor in Eq. (2.6) is the total current sourced by v_s. This current is then multiplied by a sum of two terms, one for each of the paths from the base node, contributing current into the ac ground at the output. The first term accounts for the "feedforward" portion of the current through r_f while the second term accounts for the transconductance contribution of the transistor into the ac short. In this form, we can identify factors in the response with elements in the circuit and the role that they play in the circuit operation.

The second part of the DPI approach requires finding the driving point impedance looking into the collector node. To do this we split the output node into three nodes, one for each branch as shown in Fig. 2.4b, and apply a voltage source to each of the split nodes. We find the total current supplied by the three sources when driven by identical voltages and then express the port impedance as the ratio of the applied voltage to the total current. For this operation, we set v_s, the input source magnitude, to zero, leaving its internal impedance between the input node and ground. Note that this process again removes the feedback nature of the circuit.

To identify the current components from the three sources, we initially give the sources different names, find the three currents, sum them, and then set the source voltages equal as indicated in Eq. (2.7).

$$i_{\mathrm{tot}} = v_3 \cdot g_1 + \frac{v_2}{r_{\mathrm{f}} + \frac{1}{g_{\mathrm{s}} + y_{\mathrm{ie}}}} \cdot \left(1 + \frac{g_{\mathrm{m}}}{g_{\mathrm{s}} + y_{\mathrm{ie}}}\right) + v_1 \cdot 0 \tag{2.7}$$

Source v_3 supplies a current proportional to g_l, the load conductance, the first term in Eq. (2.7). Source v_1 drives the collector node here modeled as a current source of infinite impedance, the third term in Eq. (2.7) with the zero factor.

The second term due to source v_2 is a bit more complex. The current from v_2 produces a voltage at the base node that in turn activates the transconductance source pulling current from v_1. The factor $g_m/(g_s + y_{ie})$ multiplying the current sourced by v_2 is this transconductance current. We now set the three voltage sources equal to v_x. Dividing i_{tot} by v_x and taking the reciprocal of the quotient produces the ratio v_x/i_{tot}, the DPI at the output node, Eq. (2.8). The output impedance is the reciprocal sum of the load admittance g_l and the admittance looking from the collector back to the source through r_f. This path contains a loop having the effect of reducing the output impedance by the factor $\{1 + g_m/(g_s + y_{ie})\}$ as shown in Eq. (2.8).

$$\mathrm{dpi}_{\mathrm{out}} = \frac{1}{g_1 + \frac{1}{r_{\mathrm{f}} + \frac{1}{g_{\mathrm{s}} + y_{\mathrm{ie}}}} \cdot \left(1 + \frac{g_{\mathrm{m}}}{g_{\mathrm{s}} + y_{\mathrm{ie}}}\right)} \tag{2.8}$$

The output voltage at the collector node is the product of the short-circuit current, Eq. (2.6), and the driving point impedance, Eq. (2.8). The algebra is again complex but complete to the extent that the models used to represent the elements, the transistor in particular. We simplify these relations to that shown in Eq. (2.9).

$$\begin{aligned} \mathrm{dpi}_{\mathrm{out}} &\approx \frac{r_{\mathrm{s}} + r_{\mathrm{f}}}{r_{\mathrm{s}} \cdot g_{\mathrm{m}}} \\ i_{\mathrm{sc}} &\approx \frac{-v_{\mathrm{s}} \cdot r_{\mathrm{f}} \cdot g_{\mathrm{m}}}{r_{\mathrm{s}} + r_{\mathrm{f}}} \end{aligned} \tag{2.9}$$

The output voltage is reduced to Eq. (2.10) as the classic limiting result, Eq. (2.10).

$$v_{\mathrm{out}} = i_{\mathrm{sc}} \cdot \mathrm{dpi}_{\mathrm{out}} \approx -v_{\mathrm{s}} \cdot g_{\mathrm{m}} \cdot \frac{r_{\mathrm{f}}}{r_{\mathrm{s}} g_{\mathrm{m}}} \approx -\frac{v_{\mathrm{s}} \cdot r_{\mathrm{f}}}{r_{\mathrm{s}}} \tag{2.10}$$

The method of driving point impedance is seen to divide the original circuit into two sub-circuits derived from the original to obtain the factors i_{sc} and DPI_{out}. The solution of these subproblems will naturally be easier to find than that of the whole circuit and, for single-loop designs, may many times be selected to have the feedback disabled by the process, making the analysis even easier.

Introducing Flow Graphs to the Analysis

The driving point impedance approach presented above is expanded and made systematic by adding auxiliary voltage sources to all or most of the circuit nodes and finding the short-circuit currents and driving point impedances at each of the nodes. This process partitions the problem into multiple subproblems simple enough that with practice, the DPI/SFG [2, 3] components can be written down directly by inspection and without the need of drawing a small signal equivalent circuit.

Recall the factors for the voltage division problem in Fig. 2.2. The product of these two factors can be diagramed as a flow graph, Fig. 2.5. The input signal v_s is shown driving the block containing the factor g_1. Flow graph algebra as in block diagrams has the output of a block of the product of the input arrow with the block content. The output of the first block is the short-circuit current at the output node. This arrow drives the next block containing the nodal driving point impedance. The output of the second block then is the product of the short-circuit current and DPI_{out}, producing v_{out}.

The flow graph produces a pictorial algebra implementing the DPI law, the product of the short-circuit current and the driving point impedance.

Consider next a two-loop resistive net, Fig. 2.6, with single-input source v_s. Auxiliary voltage sources v_1 and v_2 are added to the non-driven nodes tagged as (1) and (2). We find the analysis pairs, i_{scx} and DPI_x, at nodes $x = 1$ and 2 using these auxiliary voltages as temporary known entities. For node 1, we write Eq. (2.11).

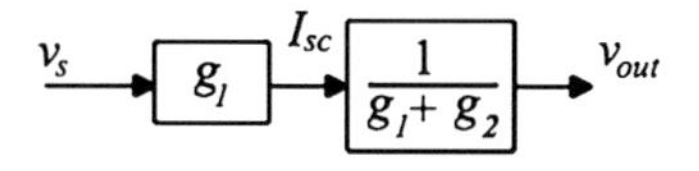

Fig. 2.5 Signal flow graph for the voltage divider problem in Fig. 2.2

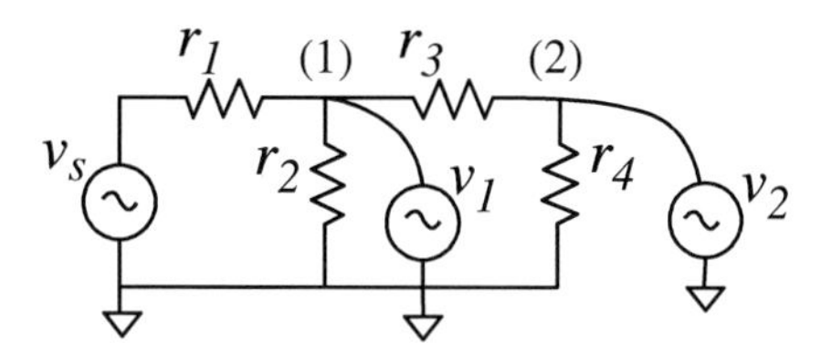

Fig. 2.6 Two-loop resistive net for DPI/SFG analysis

$$i_{sc1} = g_1 \cdot v_s + g_3 \cdot v_2$$
$$\text{dpi}_1 = \frac{1}{g_1 + g_2 + g_3} \tag{2.11}$$

The short-circuit current at node *1* is seen to have two components, one due to the source voltage through the coupling conductance g_1 and the other due to the voltage at the output node via conductance g_3. We find these terms sequentially by setting all sources but the one under consideration to zero, applying superposition. DPI_1 takes advantage of parallel conductance structures as the reciprocal of the sum of the branch conductance tied to the node.

For node 2, we write the set Eq. (2.12). Note that i_{sc2} is only due to the auxiliary voltage applied to node 1 since by superposition this node is set to zero when finding the component due to v_{in} interrupting that current path.

$$i_{sc2} = g_3 \cdot v_1$$
$$\text{dpi}_2 = \frac{1}{g_3 + g_4} \tag{2.12}$$

We now have four equations in five variable voltages, v_s, v_1, and $v_2 = v_{out}$, plus the two pairs of short-circuit currents and port impedances. The four equations are coupled through the circuit topology and through the DPI law. We can plot these as two flow graphs, one for each node, as shown in Fig. 2.7 (ignore the dashed lines for now).

The upper flow graph represents the set in Eq. (2.11), while the lower graph represents the set in Eq. (2.12), forming the DPI and short-circuit current products. The graph arrows represent either short-circuit currents or node voltages. The node voltages v_1 and v_2 are equal in both graphs, so we can couple the two graphs via these quantities, the dashed arrows in Fig. 2.7, making one graph for the circuit. This action is equivalent to eliminating variables algebraically, allowing us to find the transfer function between the input voltage source and the output node response.

Keeping the relations as shown in the coupled graphs, we can redraw the composite, untangling the whole as in Fig. 2.8.

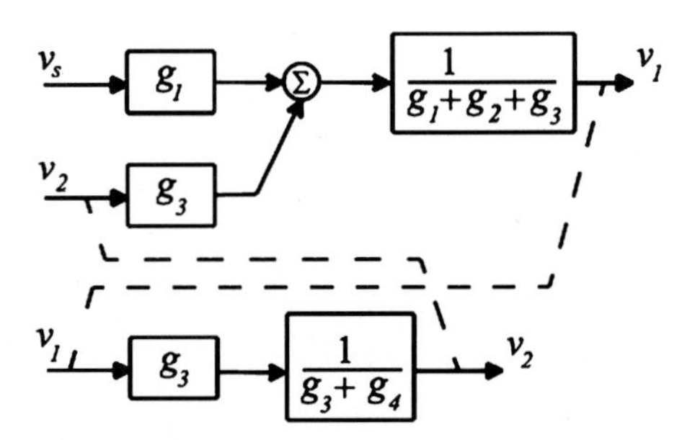

Fig. 2.7 Flow graphs for circuit in Fig. 2.6 showing coupling between graphs as *dashed arrows*

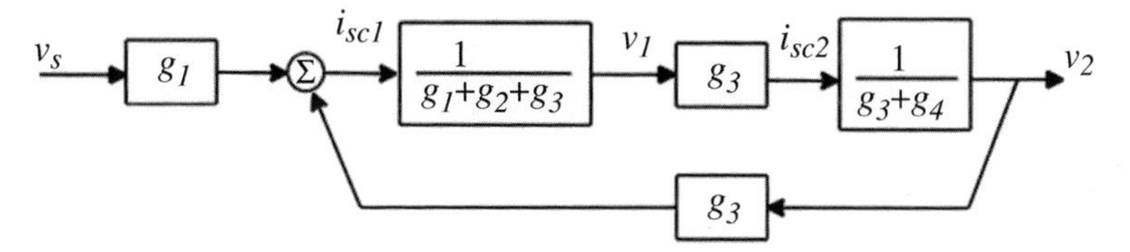

Fig. 2.8 Redrawn composite graph in Fig. 2.8 simplifying the graph

Simplified Flow Graph Algebra

To find the transfer function for the graph in Fig. 2.8, we first find the i_{sc1} arrow (or any other arrow inside the loop) as functions of v_s and $v_2(=v_{out})$. An arrow at the output of a box is the product of the input arrow and the box contents, and the output of the summer is the algebraic sum of its inputs. Then, given i_{sc1}, the $v_2(=v_{out})$ arrow is obtained as the product of i_{sc1} and product of the three blocks following i_{sc1}, factors $\{1/(g_1+g_2+g_3)\}$, g_3 and $\{1/(g_3+g_4)\}$. We then combine these results to eliminate i_{sc1} to obtain v_{out} as a function of v_s, Eq. (2.13).

$$\begin{aligned} i_{\mathrm{sc1}} &= g_1 \cdot v_{\mathrm{s}} + g_3 \cdot v_2 \\ v_2 &= i_{\mathrm{sc1}} \cdot \frac{1}{g_1+g_2+g_3} \cdot g_3 \cdot \frac{1}{g_3+g_4} \\ v_{\mathrm{out}} = v_2 &= v_{\mathrm{s}} \cdot \frac{g_1 \cdot \dfrac{1}{g_1+g_2+g_3} \cdot g_3 \cdot \dfrac{1}{g_3+g_4}}{1 - \dfrac{1}{g_1+g_2+g_3} \cdot g_3 \cdot \dfrac{1}{g_3+g_4} \cdot g_3} \end{aligned} \tag{2.13}$$

Further algebra on Eq. (2.13) can be performed to obtain a final, simplified solution.

A complete flow graph algebra exists, summarized as Mason's [2] rule. While elaborate and elegant, direct knowledge of Mason's rule is not necessary for circuit analysis as this is quickly learned by example. In the two-loop resistive net example, we found the transfer function by using basic graph definitions and eliminating variables. This process will serve to develop graph relations as needed and keep the discussion focused on the circuit analysis process. For the graph obtained in Fig. 2.8 having a single loop, the transfer function from v_s to v_2 is given as "the product of the blocks in the path between input and output arrow" divided by the quantity "1 minus the loop product," the product of the boxes in the loop. This expression is Mason's rule for a single-loop graph and will be sufficient for solving a large number of useful circuits represented by this graph structure.

It is important to note that results obtained from the graph are as complete as the modeling of the elements in the graph and of course as the completeness of the graph itself. The graph exists in the mathematical domain. *The graph is ideal.*

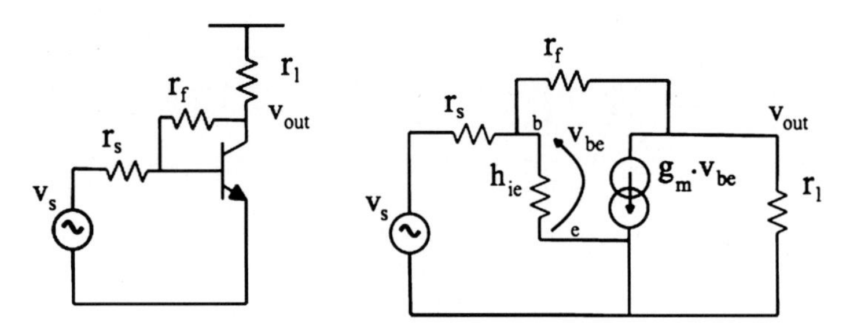

Fig. 2.9 Single transistor amplifier, repeat of Fig. 2.3

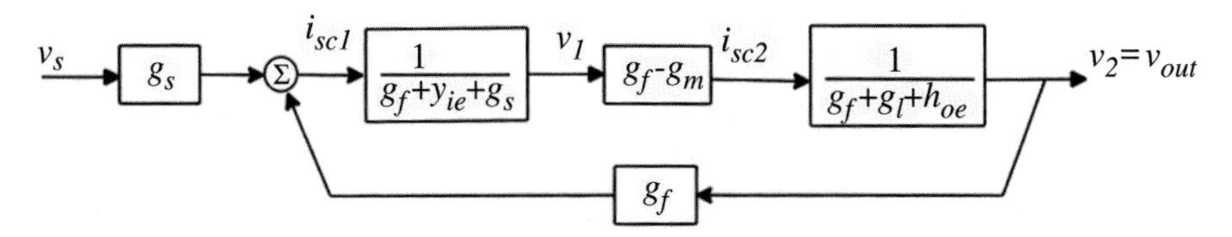

Fig. 2.10 Flow graph for feedback amplifier in Fig. 2.9

Effects sometimes forgotten in classic analysis, loading, and feedforward effects are included in the generation of the graph and appear in the final transfer functions obtained in this process automatically. The "loop product" obtained in this fashion however is NOT the feedback system loop gain. This was stated in Chap. 1 and will be fully explained in Chap. 5. It is interesting to note that the two-loop passive net with the flow graph shown in Fig. 2.8 is a feedback system as indicated by the form of the transfer function and the loop in the flow graph.

While the analysis of this passive net is interesting and served to show the DPI/SFG process, it is not very exciting. Let us apply the process to the single transistor feedback circuit of Fig. 2.3 repeated here as Fig. 2.9.

Referring to Fig. 2.9, we can write the following equation set for the base and collector nodes, the short-circuit currents, and the driving port impedances.

$$\begin{aligned} i_{\text{sc1}} &= g_{\text{s}} \cdot v_{\text{s}} + g_{\text{f}} \cdot v_{\text{out}} \\ \text{dpi}_1 &= \frac{1}{g_{\text{s}} + g_{\text{f}} + y_{\text{ie}}} \\ i_{\text{sc2}} &= g_{\text{f}} \cdot v_1 - g_{\text{m}} \cdot v_1 \\ \text{dpi}_2 &= \frac{1}{g_{\text{f}} + g_{\text{l}} + h_{\text{oe}}} \end{aligned} \tag{2.14}$$

These relations can be interpreted as the flow graph shown in Fig. 2.10.

This flow graph for the single transistor feedback amplifier has the same form as that for the two-loop passive net, Fig. 2.8. This is a basic feature of the DPI/SFG methodology—*a graph represents a circuit's topology and is valid for a large family of circuit realizations*. The process essentially creates a template extracted from the topology that is reused for a specific circuit sharing that topology.

Using the flow graph rule developed for a single-loop graph, we can write the circuit transfer function by inspection as Eq. (2.15). This relation is in a different form than that derived in Eq. (2.4) for the same circuit using nodal analysis. They are equivalent as can be seen by converting one to the other through algebraic manipulations (setting h_{oe} to 0 as assumed in Fig. 2.10). If the second term in the denominator in Eq. (2.15) is much larger than 1, we get the same limiting transfer result of ($v_{\mathrm{out}}/v_{\mathrm{s}} \sim -r_{\mathrm{f}}/r_{\mathrm{s}}$).

$$v_{\mathrm{out}} = v_{\mathrm{s}} \cdot \frac{g_{\mathrm{s}} \cdot \dfrac{1}{g_{\mathrm{s}} + g_{\mathrm{f}} + h_{\mathrm{ie}}} \cdot (g_{\mathrm{f}} - g_{\mathrm{m}}) \cdot \dfrac{1}{g_{\mathrm{f}} + g_{\mathrm{l}} + h_{\mathrm{oe}}}}{1 - \dfrac{1}{g_{\mathrm{s}} + g_{\mathrm{f}} + h_{\mathrm{ie}}} \cdot (g_{\mathrm{f}} - g_{\mathrm{m}}) \cdot \dfrac{1}{g_{\mathrm{f}} + g_{\mathrm{l}} + h_{\mathrm{oe}}}} \Rightarrow \frac{-r_{\mathrm{f}}}{r_s} \tag{2.15}$$

A Few More Examples

In this section, we show three more examples of the DPI/SFG approach to linear circuit analysis: (1) the differential input stage and the common mode/differential mode analysis, (2) the impedance of a cascode pair, and (3) finding the initial time constant for the transient response of a latching comparator.

The Differential Input Stage

The differential input circuit is usually presented as the introductory circuit for defining differential and common mode excitations and responses. The analysis uses an argument based on symmetry that allows a separation of the cell's response into two noninteracting modes—the common and the differential modes. The input signals are first defined in these terms and then applied separately to the circuit. The circuit is then analyzed using superposition for each mode separately and then the results combined the full response.

Consider the bipolar differential gain stage shown in Fig. 2.11. The input signals v_a and v_b are arbitrary.

From a strictly algebraic standpoint, we can express these input signals as equivalent common and differential mode signals as shown in Eq. (2.26).

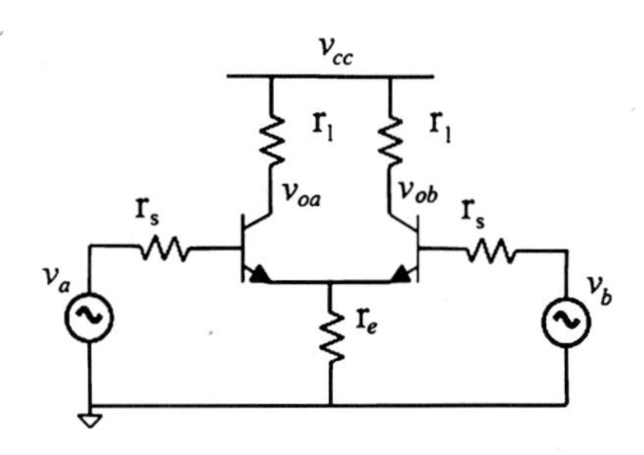

Fig. 2.11 The differential amplifier input stage

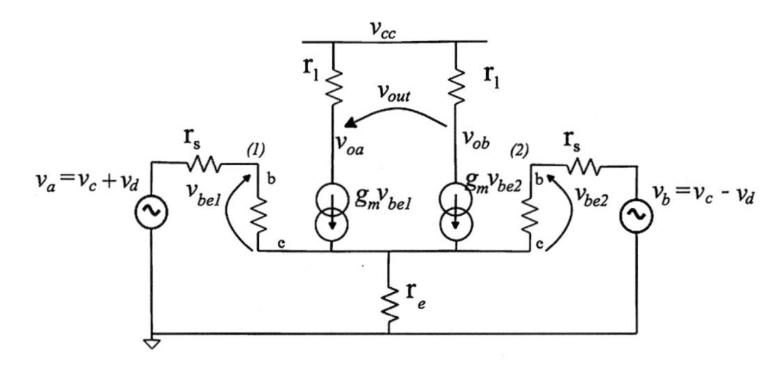

Fig. 2.12 Differential stage of Fig. 2.11 using small signal models for BJTs

$$
\begin{aligned}
v_a &= \frac{v_a}{2}+\frac{v_a}{2}+\frac{v_b}{2}-\frac{v_b}{2}=\frac{v_a-v_b}{2}+\frac{v_a+v_b}{2}=v_d+v_c \\
v_b &= \frac{v_a}{2}-\frac{v_a}{2}+\frac{v_b}{2}+\frac{v_b}{2}=-\frac{v_a-v_b}{2}+\frac{v_a+v_b}{2}=-v_d+v_c \\
v_d &= \frac{v_a-v_b}{2};\, v_c=\frac{v_a+v_b}{2}
\end{aligned}
\tag{2.16}
$$

The input signals v_a and $v_b$are now expressed in terms of sums and differences of the original signals. The term v_c, the common mode signal, is the average of the input signals, while v_d, the differential mode signal, is half of the difference signal. For common mode excitation, we set $v_a = v_b = v_c$, while for differential mode set $v_a = (-v_b) = v_d$. The net output response is then found by using superposition—find the output response for each input mode separately and then add the results for the full response. With these new equivalents for the sources, we can analyze the design using the small signal equivalent model for the bipolar transistor as shown in Fig. 2.12.

For common mode excitation, we set $v_d = 0$ (differential mode) in Fig. 2.12, making both inputs equal to the common mode value v_c. The complete symmetry of

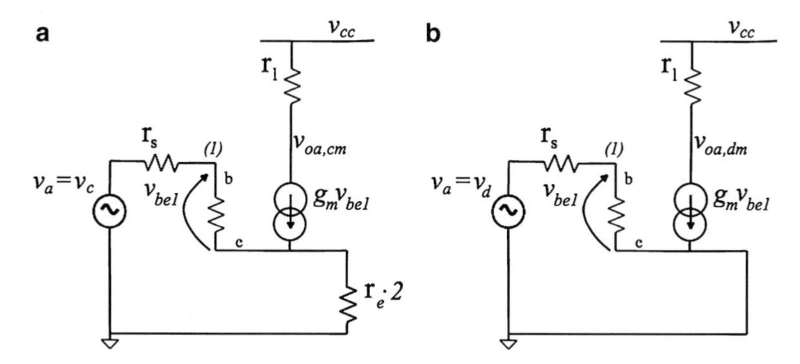

Fig. 2.13 Common mode "half-cell" (**a**) and Differential mode (**b**) for the differential amplifier stage in Fig. 2.11

the drive and circuit will make every node and current in one side equal to the corresponding voltage or current in the other. Noting that the current through the common emitter resistor r_e is twice that contributed from either side, we have a clue on how to proceed: Since this is the coupling element between left and right sides and the coupling consists of producing twice the voltage from that due to a single side we can reproduce this effect in a derived circuit with only one side (half circuit) if we double the emitter voltage response. We do this by doubling r_e as shown in Fig. 2.13a for the common mode half circuit equivalent of the differential gain cell.

This circuit is a common emitter form with emitter resistor degeneration. The impedance at the base is given in Eq. (2.17).

$$z_{\mathrm{b,cm}} = h_{\mathrm{ie}}(1 + g_{\mathrm{m}} \cdot 2 \cdot r_{\mathrm{e}}) \tag{2.17}$$

From this we can write the output voltage for the common mode half-cell as in Eq. (2.18).

$$\begin{aligned} v_{\mathrm{oa}} &= \frac{v_{\mathrm{c}}}{r_{\mathrm{s}} + h_{\mathrm{ie}}(1 + g_{\mathrm{m}} \cdot 2 \cdot r_{\mathrm{e}})} \cdot (-h_{\mathrm{ie}} g_{\mathrm{m}}) \cdot r_1 \approx \frac{v_{\mathrm{s}}}{2 \cdot r_{\mathrm{e}}} \cdot (-1) \cdot r_1 \\ v_{\mathrm{out}} &= v_{\mathrm{oa}} - v_{\mathrm{ob}} = 0 \end{aligned} \tag{2.18}$$

But node voltage v_{ob} is equal to v_{oa}, making the common mode output $(v_{oa} - v_{ob}) = 0$. The common mode voltage at both output nodes changes by the amount in Eq. (2.18) and the input impedance increases as given in Eq. (2.17).

For differential mode excitation, we set $v_c = 0$ and make the inputs $+-v_d$. Now the base nodes move in opposite direction which again from symmetry allows us to deduce a condition at the emitter node: *The effect from one source is canceled by that of the other and the emitter voltage is stationary*. The emitter node behaves like an ac ground, making the differential mode equivalent to a grounded common emitter stage. Applying an ac ground at the emitter node in Fig. 2.11 produces the differential mode half-cell in Fig. 2.13b making the input impedance and output voltages for the differential mode half-cell response as given in Eq. (2.19).

$$
\begin{aligned}
z_{\mathrm{b,dm}} &= h_{\mathrm{ie}} \\
v_{\mathrm{oa}} &= \frac{v_{\mathrm{d}}}{r_{\mathrm{s}} + h_{\mathrm{ie}}} \cdot (-h_{\mathrm{ie}} g_{\mathrm{m}}) \cdot r_1 \\
v_{\mathrm{out}} &= v_{\mathrm{oa}} - v_{\mathrm{ob}} = \frac{2 \cdot v_{\mathrm{d}}}{r_{\mathrm{s}} + h_{\mathrm{ie}}} \cdot (-h_{\mathrm{ie}} g_{\mathrm{m}}) \cdot r_1
\end{aligned}
\tag{2.19}
$$

The process used in this analysis is based on insight, experience, or in following authority. It requires one to recognize the effects of symmetry in formulating simplified equivalents. This works up to a point. For a reduced symmetry design, as is most often the case (active load for example is not symmetrical), this analysis leaves us without confidence that it fully applies. Mismatch effects are not present in this approach—we do not have a model to find the effects that we approximated out from the start by imposing complete symmetry.

Let us look at this cell using DPI/SFG. In the DPI/SFG approach, we attach auxiliary sources to the two output nodes and to the emitter node, Fig. 2.14. The DPI/SFG relations at the nodes with added auxiliary sources can be written down by inspection. We use the simplified BJT small signal model as in Fig. 2.12 to keep the analysis simple. The "short-circuit current" at the common emitter node "e" has contributions only from the voltage sources v_a and v_b as the small signal model, not using parameter h_{oe}, shows infinite impedance to this node from the output nodes v_{oa} and v_{ob}, and the driving point impedance is the parallel combination of r_e and the impedance looking into the emitters (Eq. (2.20)).

$$
\begin{aligned}
i_{\mathrm{sc,e}} &= \left(\frac{v_{\mathrm{a}}}{r_{\mathrm{s}} + h_{\mathrm{ie}}} + \frac{v_{\mathrm{a}}}{r_{\mathrm{s}} + h_{\mathrm{ie}}} \right) \cdot (1 + h_{\mathrm{ie}} g_{\mathrm{m}}) \\
\mathrm{dpi_e} &= \frac{1}{g_{\mathrm{e}} + \dfrac{2 \cdot (1 + h_{\mathrm{ie}} g_{\mathrm{m}})}{(r_{\mathrm{s}} + h_{\mathrm{ie}})}}
\end{aligned}
\tag{2.20}
$$

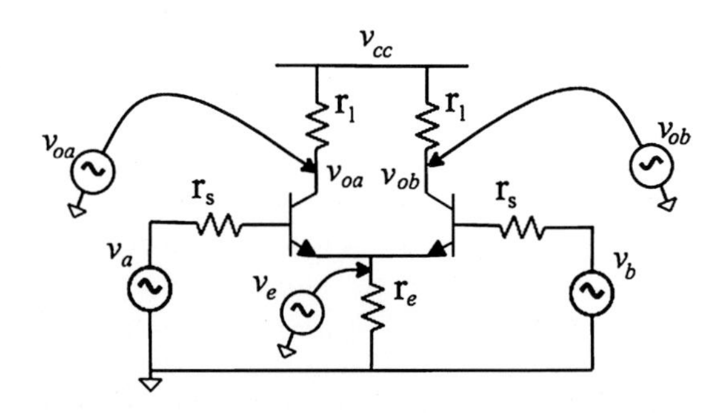

Fig. 2.14 Differential input stage with added auxiliary sources for DPI/SFG analysis

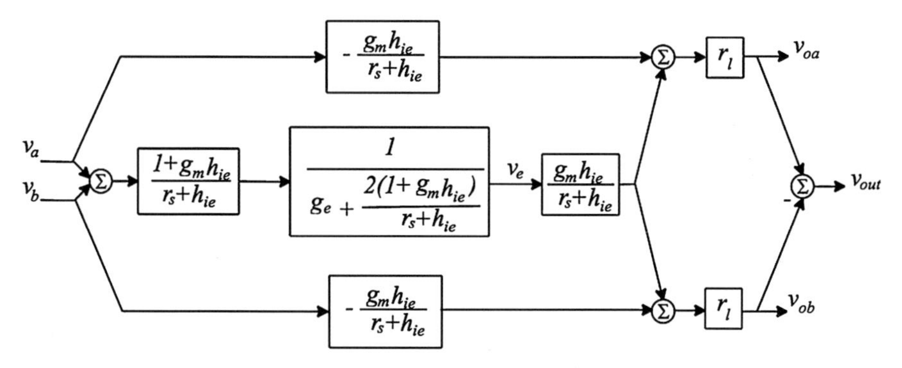

Fig. 2.15 Signal flow graph for differential input stage

The DPI relations at the output nodes, Eq. (2.21), are also written by inspection of Fig. 2.14. The short-circuit current at output v_{oa} has terms due to the input source v_a and to the auxiliary source v_e. DPI_{oa} is the parallel combination of a collector impedance ($1/h_{oe}$ assumed infinite) and the load r_l. Parameters at v_{ob} show the same form as these for v_{oa} due to the symmetry of the design. These relation sets, Eqs. (2.20) and (2.21), plot directly as the flow graph in Fig. 2.15. Combining signals entering different blocks having the same contents simplifies the flow graph.

$$
\begin{aligned}
i_{\mathrm{sc,oa}} &= \frac{v_{\mathrm{a}}}{r_{\mathrm{s}}+h_{\mathrm{ie}}}\cdot(-h_{\mathrm{ie}}g_{\mathrm{m}})+\frac{v_{\mathrm{e}}}{r_{\mathrm{s}}+h_{\mathrm{ie}}}\cdot(h_{\mathrm{ie}}g_{\mathrm{m}})\\
\mathrm{dpi}_{\mathrm{oa}} &= r_{\mathrm{la}}\\
i_{\mathrm{sc,ob}} &= \frac{v_{\mathrm{b}}}{r_{\mathrm{s}}+h_{\mathrm{ie}}}\cdot(-h_{\mathrm{ie}}g_{\mathrm{m}})+\frac{v_{\mathrm{e}}}{r_{\mathrm{s}}+h_{\mathrm{ie}}}\cdot(h_{\mathrm{ie}}g_{\mathrm{m}})\\
dpi_{\mathrm{ob}} &= r_{\mathrm{lb}}
\end{aligned}
\tag{2.21}
$$

There are no loops in this graph, making the graph algebra simply the sum of the paths. The output signals (with $r_{la}=r_{lb}=r_l$) become

$$
\begin{aligned}
v_{\mathrm{oa}} &= v_{\mathrm{a}}\frac{-g_{\mathrm{m}}h_{\mathrm{ie}}}{r_{\mathrm{s}}+h_{\mathrm{ie}}}\cdot r_{\mathrm{l}}+(v_{\mathrm{a}}+v_{\mathrm{b}})\cdot\frac{1+h_{\mathrm{ie}}g_{\mathrm{m}}}{r_{\mathrm{s}}+h_{\mathrm{ie}}}\cdot\frac{1}{g_{\mathrm{e}}+\dfrac{2\cdot(1+h_{\mathrm{ie}}g_{\mathrm{m}})}{(r_{\mathrm{s}}+h_{\mathrm{ie}})}}\cdot\frac{h_{\mathrm{ie}}g_{\mathrm{m}}}{\dfrac{v_{\mathrm{e}}}{r_{\mathrm{s}}+h_{\mathrm{ie}}}}\cdot r_{\mathrm{l}}\\
v_{\mathrm{ob}} &= v_{\mathrm{b}}\frac{-g_{\mathrm{m}}h_{\mathrm{ie}}}{r_{\mathrm{s}}+h_{\mathrm{ie}}}\cdot r_{\mathrm{l}}+(v_{\mathrm{a}}+v_{\mathrm{b}})\cdot\frac{1+h_{\mathrm{ie}}g_{\mathrm{m}}}{r_{\mathrm{s}}+h_{\mathrm{ie}}}\cdot\frac{1}{g_{\mathrm{e}}+\dfrac{2\cdot(1+h_{\mathrm{ie}}g_{\mathrm{m}})}{(r_{\mathrm{s}}+h_{\mathrm{ie}})}}\cdot\frac{h_{\mathrm{ie}}g_{\mathrm{m}}}{\dfrac{v_{\mathrm{e}}}{r_{\mathrm{s}}+h_{\mathrm{ie}}}}\cdot r_{\mathrm{l}}\\
v_{\mathrm{out}} &= (v_{\mathrm{a}}-v_{\mathrm{b}})\cdot\frac{-g_{\mathrm{m}}h_{\mathrm{ie}}}{r_{\mathrm{s}}+h_{\mathrm{ie}}}\cdot r_{\mathrm{l}}
\end{aligned}
\tag{2.22}
$$

The differential output is as before, $v_{out}=v_{oa}-v_{ob}$. Taking this difference causes the terms multiplied by the factor (v_a+v_b) to cancel (the common mode gain is 0), leaving the term multiplied by the factor $(v_a - v_b)$, the difference mode, as we saw using the half-cell representations. In the DPI/SFG analysis, we did not need this insight or experience, and symmetry of design or elements was not imposed.

Under the assumption of complete symmetry, the signal flow graph and the half-cell analysis give the same results as required. A marked difference is that the flow graph is derived systematically without the insight required in the half-cell formulations—Eq. (2.22) was obtained without imposing any conditions on the input voltages. Further, the common mode excitation result of zero common mode gain is due to the complete matching between corresponding elements on each side so that each half of the circuit processes the common mode signal exactly the same. Asymmetry is not treated in the half-cell formulation, while any mismatch of corresponding elements can be tracked in the DPI/SFG approach. It is easy to insert a different value for r_{la} than that for r_{lb} in the flow graph in Fig. 2.15, for example, so that the resulting solutions for both common and differential mode excitation reflect this mismatch in elements properly with no additional effort. Similarly, mismatch in transistor and other element parameters can be included for a full analysis of real circuits having variation in components.

The Cascode Configuration

The cascode configuration using nMOS transistors is shown in Fig. 2.16. This configuration of stacked active devices has good high-frequency response and large voltage gain from v_{in} to v_{out}. Here we solve for the impedance [4], looking into the drain of Q_2 using only the DPI/SFG approach.

First consider the response of this circuit to a small signal current excitation at the output node, i_{test}. Using the simple transistor model consisting of transconductance generator g_m and output impedance r_{ds} only, we argue that this current will increase v_{out} which then increases the voltage at node *1* due to coupling through r_{ds} of Q_1. This node *1* voltage produces a return current to the v_{out} node, further increasing v_{out}. The circuit shows positive feedback with the loop internal to device

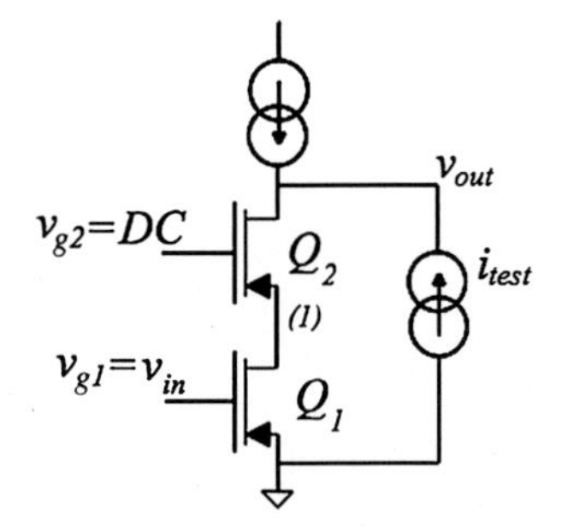

Fig. 2.16 nMOS cascode circuit for high output impedance

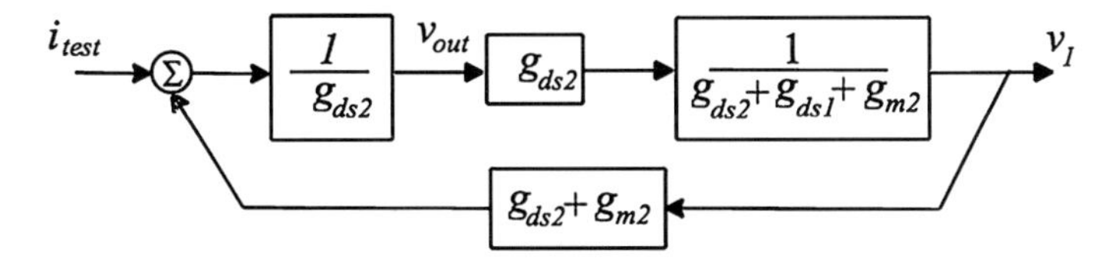

Fig. 2.17 Signal flow graph for cascode impedance circuit shown in Fig. 2.16

Q_1. This argument is not required for the analysis but is interesting to note. We find the DPI relations at both nodes by inspection Eq. (2.23):

$$\begin{aligned} i_{\text{sc,out}} &= i_{\text{test}} + v_1 \cdot (g_{\text{m2}} + g_{\text{ds2}}) \\ \text{dpi}_{\text{out}} &= \frac{1}{g_{\text{ds2}}} \\ i_{\text{sc1}} &= v_{\text{out}} \cdot g_{\text{ds2}} \\ \text{dpi}_{\text{out}} &= \frac{1}{g_{\text{ds1}} + g_{\text{ds2}} + g_{\text{m2}}} \end{aligned} \tag{2.23}$$

These relations plot onto the flow graph shown in Fig. 2.17. To find the output impedance, we find the voltage at the output node due to i_{test} and then take the ratio $v_{out}/i_{test} = z_{out}$.

$$\begin{aligned} z_{\text{out}} = \frac{v_{\text{out}}}{i_{\text{test}}} &= \frac{\frac{1}{g_{\text{ds2}}}}{1 - \frac{g_{\text{m2}} + g_{\text{ds2}}}{g_{\text{ds1}} + g_{\text{ds2}} + g_{\text{m2}}}} \\ &= r_{\text{ds1}} + r_{\text{ds2}} + g_{\text{m2}} r_{\text{ds1}} r_{\text{ds2}} \end{aligned} \tag{2.24}$$

This last result shows that the cascode output impedance is approximately the product of the intrinsic gain of the upper device ($r_{ds2} \times g_{m2}$) and the output impedance of the lower device, r_{ds1}, which we argued is due to the positive feedback structure.

Initial Time Constant Response for a Reset Comparator

Comparators are high gain, high-performance amplifiers used in analog-to-digital converters (ADCs) and other applications where the input signal is compared to a reference and the output is to be the digital representation of the sign of the difference between the input voltage and the reference, $V_{in} - V_{ref}$. An important design parameter is the small signal response time when the input signal is close to the reference value. If an input signal is very close to the reference, the comparator

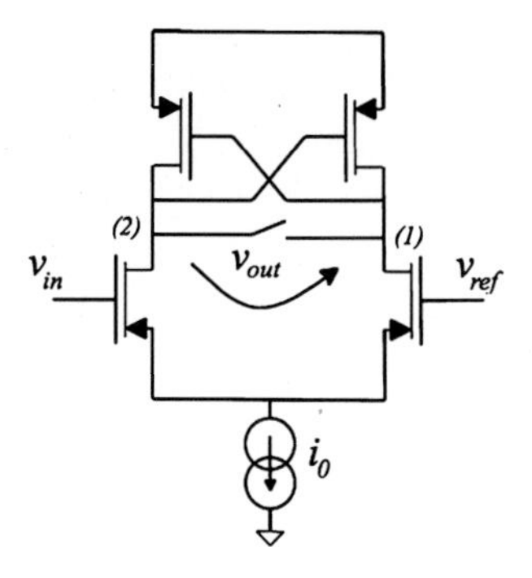

Fig. 2.18 Circuit for reset comparator input stage

response will be sluggish, limiting the useful quantization clock rate attained in the ADC, the conversion rate.

Consider the comparator input stage shown in Fig. 2.18. This circuit operates in two modes: reset and latch. During the reset mode, the switch between nodes 1 and 2 is closed, and the pMOS load transistors operate as diode-connected loads in parallel. Nodes *1* and *2* are at $V_{dd} - V_{gs}(I_0/2)$ since half of the tail current flows through each of the diode-connected transistors. When the switch is opened, the output nodes are released, and the pMOS transistors provide positive feedback, amplifying the difference in input signals regeneratively (positive feedback). The response for short time is exponential, making the time to attain sufficient "detection" amplitude for the following stage dependent on the initial input difference and a system time constant. We wish to solve for the initial response time constants, describing the transient as we release the latch from reset by opening the switch [5].

For V_{in} close to V_{ref}, we can define the inputs as the differential signal $v_d = + -(V_{in} - V_{Ref})/2$ and replace the differential input transistors with source currents $+ -v_d \times g_m$ into the drains of the cross-coupled pMOS devices as shown in Fig. 2.19. This simplifies the next step of writing the DPI relations at nodes *1* and *2* (Eq. (2.25)) just after the switch opens.

$$\begin{aligned} i_{\text{sc1}} &= v_{\text{d}} \cdot g_{\text{mn}} - v_2 \cdot g_{\text{mp}} \\ \text{dpi}_1 &= \frac{1}{g_{\text{dsn}} + g_{\text{dsp}} + cs} = \frac{r}{1 + rcs}; r = \frac{1}{g_{\text{dsn}} + g_{\text{dsp}}} \\ i_{\text{sc2}} &= -v_{\text{d}} \cdot g_{\text{mn}} - v_1 \cdot g_{\text{mp}} \\ \text{dpi}_2 &= \frac{1}{g_{\text{dsn}} + g_{\text{dsp}} + cs} = \frac{r}{1 + rcs} \end{aligned} \tag{2.25}$$

The capacitance "*c*" in Eq. (2.25) is the total capacitance at each output node due to the two MOS drains and one pMOS gate, and "*r*" is the parallel combination of the two drain-source conductance $g = (g_{dsn} + g_{dsp}) = 1/r$. Note that these relations can

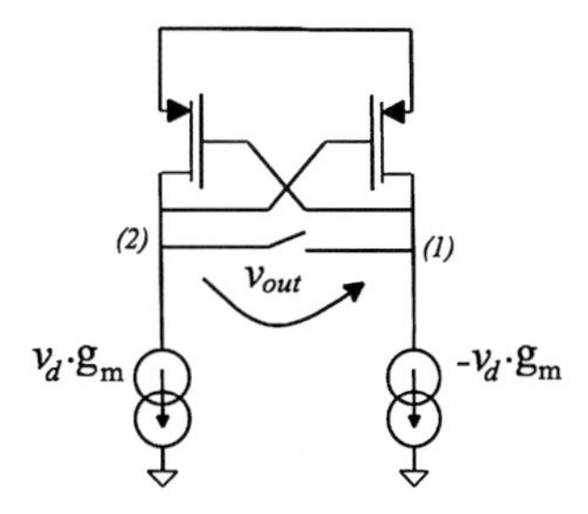

Fig. 2.19 Reset comparator from Fig. 2.18 with input diff pair replaced by small signal current sources

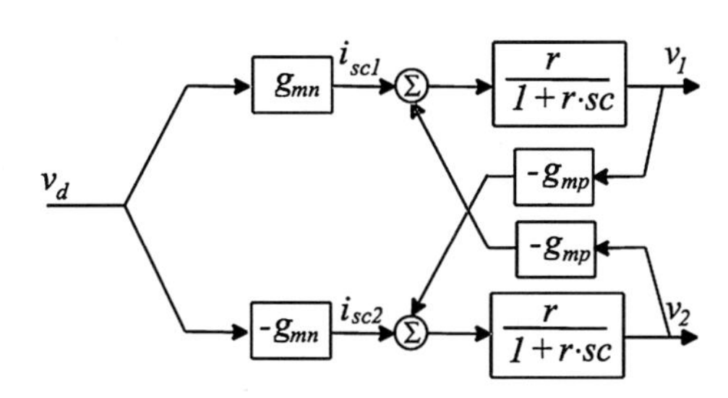

Fig. 2.20 Signal flow graph for reset comparator of Fig. 2.19

be written down by inspection. We interpret Eq. (2.25) as the system flow graph shown as Fig. 2.20.

The system transfer function $H(s)$ is now obtained by solving the flow graph for $(v_1 - v_2)$ as a function of v_d. There are two paths to each of the output nodes and there is one loop. Solving for both outputs, we get Eq. (2.6).

$$\begin{aligned}
v_1 &= \frac{v_\mathrm{d} \cdot g_\mathrm{mn} \cdot \dfrac{r}{1+rcs} \cdot \left(1 + g_\mathrm{mp} \cdot \dfrac{r}{1+rcs}\right)}{1 - \left(g_\mathrm{mp} \cdot \dfrac{r}{1+rcs}\right)^2} \\
v_2 &= \frac{-v_\mathrm{d} \cdot g_\mathrm{mn} \cdot \dfrac{r}{1+rcs} \cdot \left(1 - g_\mathrm{mp} \cdot \dfrac{r}{1+rcs}\right)}{1 - \left(g_\mathrm{mp} \cdot \dfrac{r}{1+rcs}\right)^2} \\
H(s) &= \frac{v_1 - v_2}{v_\mathrm{d}} = \frac{2 \cdot g_\mathrm{mn} \cdot \dfrac{r}{1+rcs}}{1 - \left(g_\mathrm{mp} \cdot \dfrac{r}{1+rcs}\right)^2}
\end{aligned} \tag{2.26}$$

The system characteristic frequencies are found by setting the denominator of $H(s)$, Eq. (2.26), to zero and solving for the roots of "s":

$$\begin{aligned}
1-\left(g_{\mathrm{mp}}\cdot\tfrac{r}{1+rcs}\right)^2 &= 0\\
\left(g_{\mathrm{mp}}\cdot\frac{r}{1+rcs}\right) &= \pm 1\\
s=\frac{-1\pm g_{\mathrm{mp}}\cdot r}{rc} &= s=\frac{-1\pm A_0}{\tau}\\
A_0 &= g_{\mathrm{mp}}\cdot r\\
\tau &= rc
\end{aligned} \tag{2.27}$$

where $\tau = rc = c/(g_{dsp} + g_{dsn})$ and $A_O = g_{mp}r$.

The reset comparator, a second-order system, has two poles, one in the left half plane and one in the right. It is the pole in the right half plane that produces the growing exponential due to positive feedback (Eq. (2.28)).

$$\tau_{\mathrm{eff}} = \frac{-1+A_0}{\tau} \approx \frac{g_{\mathrm{mp}}}{c} \tag{2.28}$$

In addition to this final result of finding the positive time constant on opening the reset switch, the DPI/SFG process has produced the full transfer function (Eq. (2.26)) and both system poles (Eq. (2.27)) directly. Also note that while the system has positive feedback, this fact did not cause additional complexity in the process as the process used sub-circuits where the feedback was deactivated.

Summary

The DPI approach to circuit analysis has been presented as a systematic application of superposition plus the use of auxiliary supplies, producing nodal responses as a simple product of a short-circuit current and the driving point impedance, $I_{sc} \times DPI$, at the desired output node. This approach was seen to divide the initial circuit into sub-circuit problems, circuits that are easier to solve.

In addition to the reduction in complexity, the process provided another benefit of removing feedback in generated sub-circuits. This process led naturally to the use of multiple auxiliary sources and not just at the desired output node, producing a system of coupled sub-circuits. These coupled sub-circuits were in turn represented graphically as a signal flow graph where simple graph algebra is used to produce the desired transfer relations.

Even with the limited examples presented in this chapter, a pattern was seen in the procedure and in resulting flow graphs: All feedback circuits having a single input and two internal nodes produce the same graph, the graph being a property of the circuit topology. Resistive nets, filters, and feedback amplifiers all map onto the same graph form consisting of a single loop.

Once a circuit is mapped onto a flow graph, it is necessary to do graph algebra to extract the desired transfer functions. With only the simple defining relations of block input–output, enough graph algebra was developed to be able to solve these graphs.

Armed with the methodology and a basic graph reduction knowledge, we were then able to solve a few interesting circuits. Beyond the resistive nets and dividers, a single transistor feedback stage was solved using DPI only and then again using DPI/SFG. *Feedback was not treated directly but was handled automatically by the procedure, the key feature explored in this text.*

The next circuit presented was the differential input stage that was solved without the need of the half-cell formulation of common mode/differential mode excitation generally used. This example is instructive in that the conditions of applicability of the half-cells and differential/common mode excitation arise in the algebra produced in the process. Symmetry assumed a priori using half-cells is now seen to produce factors that become zero when symmetry is imposed. Since the DPI/SFG result did not require symmetry, these results are complete for the nonsymmetrical case and allow for mismatch analysis not available when using the symmetric-defined half-cell models.

Another feedback circuit, the output impedance for a cascode configuration, was then presented. Again no particular notice was given to the feedback configuration, which in this case was one of positive feedback.

The final example was finding the initial linear response time constant for a reset comparator where we directly found the full transfer function and both characteristic time constants for the second-order system, one producing regenerative positive feedback and the other negative. Again the feedback in the system did not complicate the analysis.

While we began with trivial resistive nets as development vehicles, the chapter demonstrated the DPI/SFG methodology with some interesting and nontrivial circuits to further develop the appeal for this approach. In the succeeding chapters, the methodology will be further developed and generalized into a general purpose and useful design tool for the analog designer.

References

1. R. Kelly, Electronic circuit analysis and design by driving-point impedance techniques. IEEE Trans. Edu. **13**(3), 153–167 (1970)
2. S.J. Mason, Feedback theory—some properties of signal flow graphs. Proc. IRE **64**, 1144–1156 (1953)

3. A. Ochoa, A systematic approach to the analysis of general and feedback circuits and systems using signal flow graphs and driving-point impedance. IEEE TCAS-II **45**(2), 187–195 (1998)
4. A. Abide, On the operation of cascode gain stages. IEEE J. Solid-St. Circ. **23**(6), 1434–1437 (1988)
5. D. Johns, K. Martin, *Analog Integrated Circuit Design* (John Wiley & Sons, New York, NY, 1997)

Chapter 3
DPI/SFG Generalizations and Expansions: The Road to Feedback

Abstract The dpi/sfg methodology is a systematic analysis procedure for analog electronic circuits. It provides a process that interprets or maps a circuit onto a flow graph without the use of small signal equivalent circuits, followed by graph algebra and graph manipulations and transformations to generate input/output transfer relations. The flow graph reflects the circuit topology so that many circuits map onto the same flow graph form—the block contents are different reflecting the particular circuit properties. This similarity of form allows us to generalize terminology and to create template flow graphs and transfer function responses. Additionally, we are then able to extend the general approach to include fully embedded linear blocks, both classic and blocks that transform a signal into another domain, into the methodology.

In Chap. 2 we noted that many circuits map onto the same flow graph template, indicating that a graph represents the form or topology of the circuit as well as the circuit itself. We saw this commonality in graphs for the two-loop resistive net, the single transistor feedback amp, and the cascode impedance circuit. This single-loop graph is in fact common to all circuits having *or modeled as having* an input and two additional nodes with feedback between these two nodes. The graph represents a higher-level abstraction. In this chapter, we will define graph parameters to emphasize this commonality and to explore generalizations such terminology allows. This process will show that we can extend the application of DPI/SFG to include sub-cells, first classic linear and then to include transducer blocks where the signal changes form, but a linear relation is maintained.

Flow Graph Components

We return to the two-loop resistive net presented in Fig. 2.6, shown in Fig. 3.1, to begin this discussion on generalizations of loop analysis.

We write the defining DPI relations by inspection, Eqs. (3.1) and (3.2), and plot these as the circuit flow graph in Fig. 3.2.

A. Ochoa, *Feedback in Analog Circuits*, DOI 10.1007/978-3-319-26252-9_3

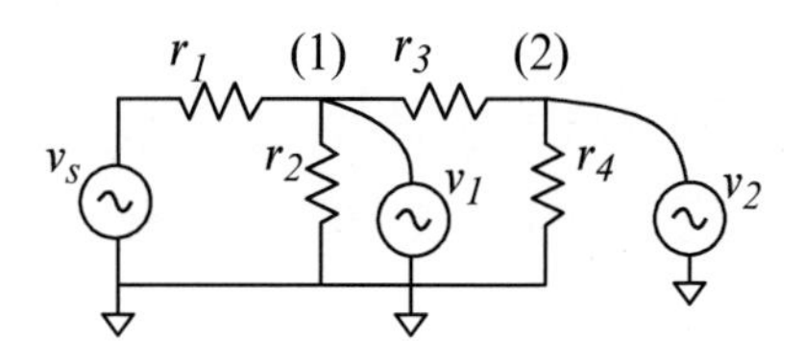

Fig. 3.1 The resistive two-loop net circuit

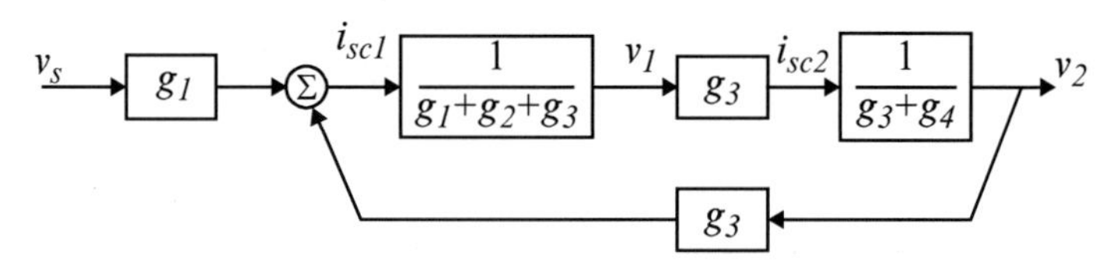

Fig. 3.2 Signal flow graph for the two-loop resistive net in Fig. 3.1

$$
\begin{aligned}
i_{\mathrm{sc1}} &= g_1 \cdot v_{\mathrm{s}} + g_3 \cdot v_2 \\
\mathrm{dpi}_1 &= \frac{1}{g_1 + g_2 + g_3}
\end{aligned}
\tag{3.1}
$$

$$
\begin{aligned}
i_{\mathrm{sc2}} &= g_3 \cdot v_1 \\
\mathrm{dpi}_2 &= \frac{1}{g_3 + g_4}
\end{aligned}
\tag{3.2}
$$

The procedure we followed in generating this graph, defining nodal short-circuit currents and driving point impedances, results in a consistent set of relations that when interpreted as a flow graph reveals the circuit topology as a common graph for a family of circuits. To fully see this, we define a universal terminology that helps simplify the discussions and generalize the results.

In Eq. (3.1), the current into the ac short applied to node 1 has two components, one for each of the adjacent nodes. Each current is the product of a node voltage and the "conductance" coupling the drive node to the ac short node. In the graphical flow graph representation, these components are shown as ideal unidirectional blocks, a current into one node due to a voltage at another. The coupling "conductance" becomes a "transconductance" element in the graph.

We can represent these coupling transconductances as g_{mij} with the index "i" as the receiving node and "j" the sourcing node. i_{sc1} and i_{sc2} become Eq. (3.3).

$$
\begin{aligned}
i_{\mathrm{sc1}} &= g_{\mathrm{m1s}} \cdot v_{\mathrm{s}} + g_{\mathrm{m12}} \cdot v_2 \\
i_{\mathrm{sc2}} &= g_{\mathrm{m21}} \cdot v_1
\end{aligned}
\tag{3.3}
$$

If we rename the driving point impedance for node "j" z_j, we further simplify the notation as shown in the following equivalent circuit representation in Fig. 3.3 and the flow graph in Fig. 3.4.

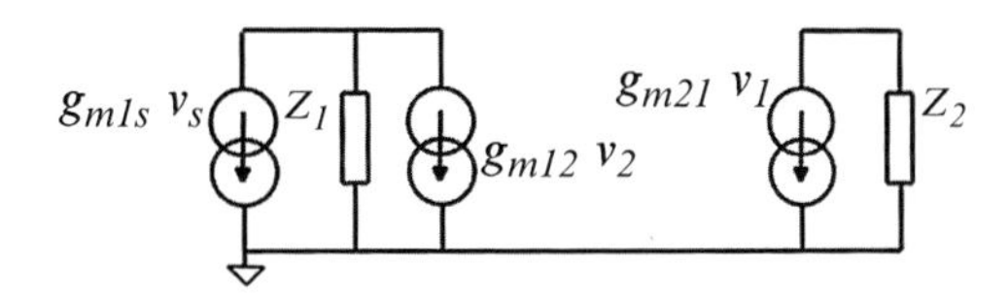

Fig. 3.3 Equivalent small signal representation for Equation set (3.3)

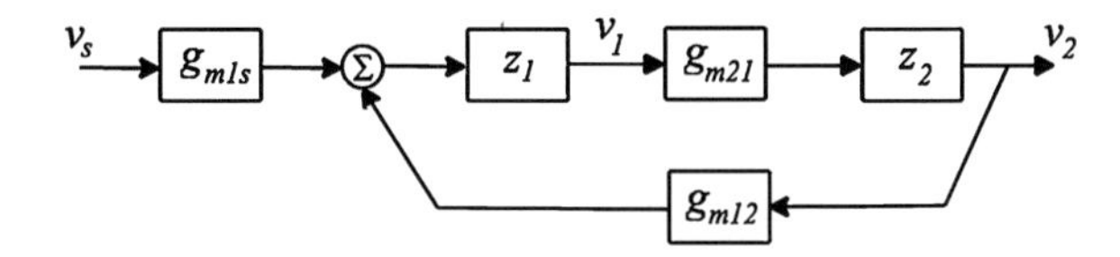

Fig. 3.4 Signal flow graph for circuit in Fig. 3.1 in universal notation

The Flow Graph Template

The flow graph in Fig. 3.4 is the parameterized version of the graph in Fig. 3.2 that was generated for the two-loop resistive net. Solving the flow graph for the transfer functions to nodes 1 and 2, we get the voltage transfer relations for the two-loop topology, Eq. (3.4), to each of the two non-driven nodes:

$$
\begin{aligned}
H_1 &= \frac{g_{\mathrm{m1s}} \cdot z_1}{1 - z_1 \cdot g_{\mathrm{m21}} \cdot z_2 \cdot g_{\mathrm{m12}}} \\
H_2 &= \frac{g_{\mathrm{m1s}} \cdot z_1 \cdot g_{\mathrm{m21}} \cdot z_2}{1 - z_1 \cdot g_{\mathrm{m21}} \cdot z_2 \cdot g_{\mathrm{m12}}}
\end{aligned}
\tag{3.4}
$$

Specific parameters for the two-loop resistive net circuit in Fig. 3.1 are

$$
\begin{aligned}
&g_{\mathrm{m1s}} = g_1; g_{\mathrm{m21}} = g_{\mathrm{m12}} = g_3 \\
&z_1 = \frac{1}{g_1 + g_2 + g_3}; z_2 = \frac{1}{g_3 + g_4}
\end{aligned}
\tag{3.5}
$$

Substituting the circuit-specific parameters for the two-loop resistive net, Eq. (3.5), into the template in Eq. (3.4), we find the transfer functions H_1 and H_2 as shown in Eq. (3.6):

$$
\begin{aligned}
H_1 &= \frac{g_1 \cdot \dfrac{1}{g_1 + g_2 + g_3}}{1 - \dfrac{1}{g_1 + g_2 + g_3} \cdot g_3 \cdot \dfrac{1}{g_1 + g_2 + g_3} \cdot g_3} \\
H_1 &= \frac{g_1 \cdot \dfrac{1}{g_1 + g_2 + g_3} \cdot g_3 \cdot \dfrac{1}{g_3 + g_4}}{1 - \dfrac{1}{g_1 + g_2 + g_3} \cdot g_3 \cdot \dfrac{1}{g_3 + g_4} \cdot g_3}
\end{aligned}
\tag{3.6}
$$

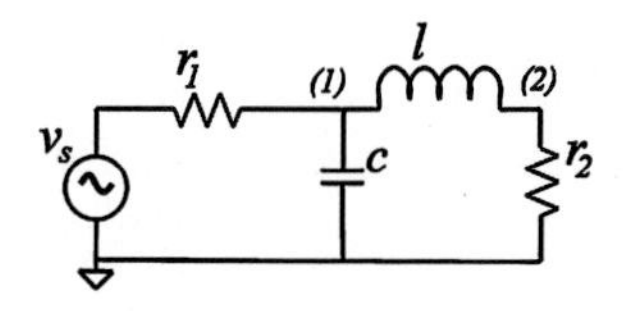

Fig. 3.5 Modified 2-loop passive circuit

We reuse this graph and template response functions for other circuits by defining the graph parameters specific to the new circuit. The circuit shown in Fig. 3.5, for example, is mapped onto this graph by replacing the general parameters with those given in Eq. (3.7), resulting in the transfer functions in Eq. (3.8).

$$
\begin{aligned}
g_{\mathrm{m1s}} &= g_1; g_{\mathrm{m21}} = g_{\mathrm{m12}} = \frac{1}{ls} \\
z_1 &= \frac{1}{g_1 + cs + \frac{1}{ls}}; z_2 = \frac{1}{g_2 + \frac{1}{ls}}
\end{aligned}
\tag{3.7}
$$

$$
\begin{aligned}
H_1 &= \frac{g_1 \cdot \frac{1}{g_1 + cs + \frac{1}{ls}}}{1 - \frac{1}{g_1 + cs + \frac{1}{ls}} \cdot \frac{1}{ls} \cdot \frac{1}{g_2 + \frac{1}{ls}} \cdot \frac{1}{ls}} \\
H_2 &= \frac{g_1 \cdot \frac{1}{g_1 + cs + \frac{1}{ls}} \cdot \frac{1}{ls} \cdot \frac{1}{g_2 + \frac{1}{ls}}}{1 - \frac{1}{g_1 + cs + \frac{1}{ls}} \cdot \frac{1}{ls} \cdot \frac{1}{g_2 + \frac{1}{ls}} \cdot \frac{1}{ls}}
\end{aligned}
\tag{3.8}
$$

Specific circuit elements for this *r*–*l*–c filter can now be used to define the parameterized flow graph elements and entered into the transfer function template relations in Eq. (3.8). With $r_1 = 100\ k\Omega$, $c = 1\ pF$, $l = 10\ mH$, and $r_2 = 100\ k\Omega$, we get the Bode responses shown in Fig. 3.6, H_1 to node *1* and H_2 to node *2*. H_2 shows a second-order frequency roll-off, while H_1 shows a first-order response.

Circuits with active elements are handled in the same manner. The circuit parameters for the bipolar feedback amplifier shown in Fig. 3.7 translate to flow graph variables given in Eq. (3.9).

$$
\begin{aligned}
g_{\mathrm{m1s}} &= g_{\mathrm{s}}; g_{\mathrm{m21}} = (-g_{\mathrm{m}} + g_{\mathrm{f}} + cs); g_{\mathrm{m12}} = g_{\mathrm{f}} + cs \\
z_1 &= \frac{1}{g_{\mathrm{s}} + g_{\mathrm{f}} + y_{\mathrm{ie}} + cs}; z_2 = \frac{1}{g_1 + g_{\mathrm{f}} + cs}
\end{aligned}
\tag{3.9}
$$

Substituting these parameters into our two-node template in Eq. (3.4), we get the transfer functions H_1 to the bipolar base node and H_2 to the collector node, Eq. (3.10).

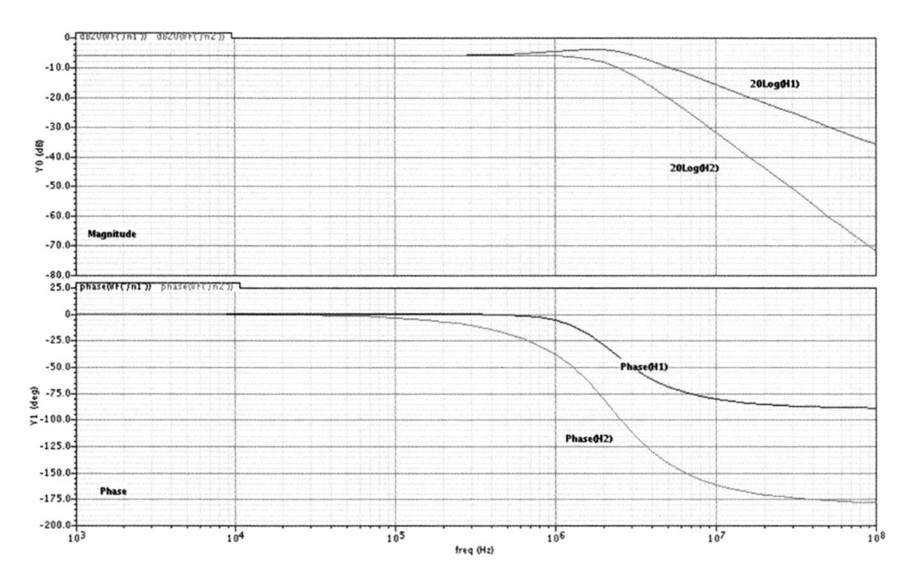

Fig. 3.6 Magnitude and phase bode plots for H_1 and H_2 for filter circuit in Fig. 3.5

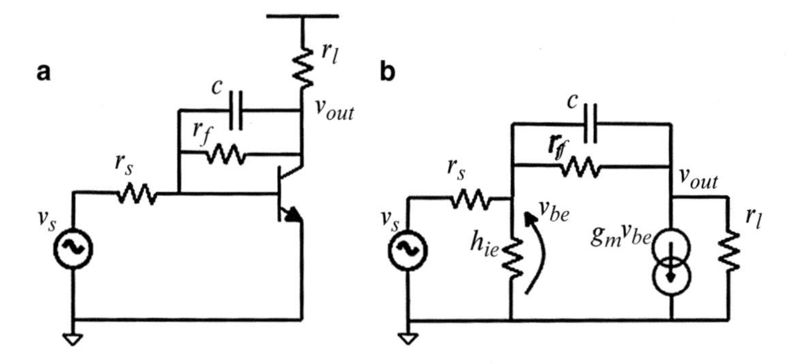

Fig. 3.7 (**a**) A bipolar feedback amplifier and (**b**) its small signal representation

$$H_1 = \frac{g_s \cdot \dfrac{1}{g_s + g_f + y_{ie} + cs}}{1 - \dfrac{(-g_m + g_f + cs)}{g_s + g_f + y_{ie} + cs} \cdot \dfrac{(g_f + cs)}{g_l + g_f + cs}}$$

$$H_2 = \frac{g_s \cdot \dfrac{(-g_m + g_f + cs)}{g_s + g_f + y_{ie} + cs} \cdot \dfrac{1}{g_l + g_f + cs}}{1 - \dfrac{(-g_m + g_f + cs)\cdot}{g_s + g_f + y_{ie} + cs} \cdot \dfrac{(g_f + cs)}{g_l + g_f + cs}} \tag{3.10}$$

While Eq. (3.10) is complex, let us look at what this expression approaches at low frequency for H_2, and at "high loop product." Combining the factors common to both numerator and denominator, we rewrite the expression for H_2 in Eq. (3.10) as shown in Eq. (3.11). If the loop product, the expression subtracted from 1 in the denominator, is much greater than *1*, we get that the transfer function H_2 is defined as the negative of the ratio of the feedback resistor r_f to the source resistor r_s:

$$
\begin{aligned}
H_2 &= \frac{g_\mathrm{s} \cdot \alpha \cdot \dfrac{1}{g_\mathrm{l} + g_\mathrm{f} + cs}}{1 - \alpha \cdot \dfrac{1}{g_\mathrm{l} + g_\mathrm{f} + cs} \cdot (g_\mathrm{f} + cs)} \\
\alpha &= \frac{1}{g_\mathrm{s} + g_\mathrm{f} + + y_\mathrm{ie} cs} \cdot (-g_\mathrm{m} + g_\mathrm{f} + cs) \\
H_2(\mathrm{lowFreq}) &\approx \frac{-g_\mathrm{s}}{g_\mathrm{f}} = \frac{-r_\mathrm{f}}{r_\mathrm{s}}
\end{aligned}
\tag{3.11}
$$

Simulation results for this circuit using $r_s = 10\ k\Omega$, $c = 1\ pF$, $r_f = 100\ k\Omega$, and $r_l = 1\ M\Omega$ are shown in the Fig. 3.8 for the magnitude of the voltage transfer function (top plot) and its phase (bottom). Note that the low-frequency magnitude is 19.22 dB, slightly less than the 20 dB predicted by Eq. (3.11), due to the loop product not being much larger than 1.

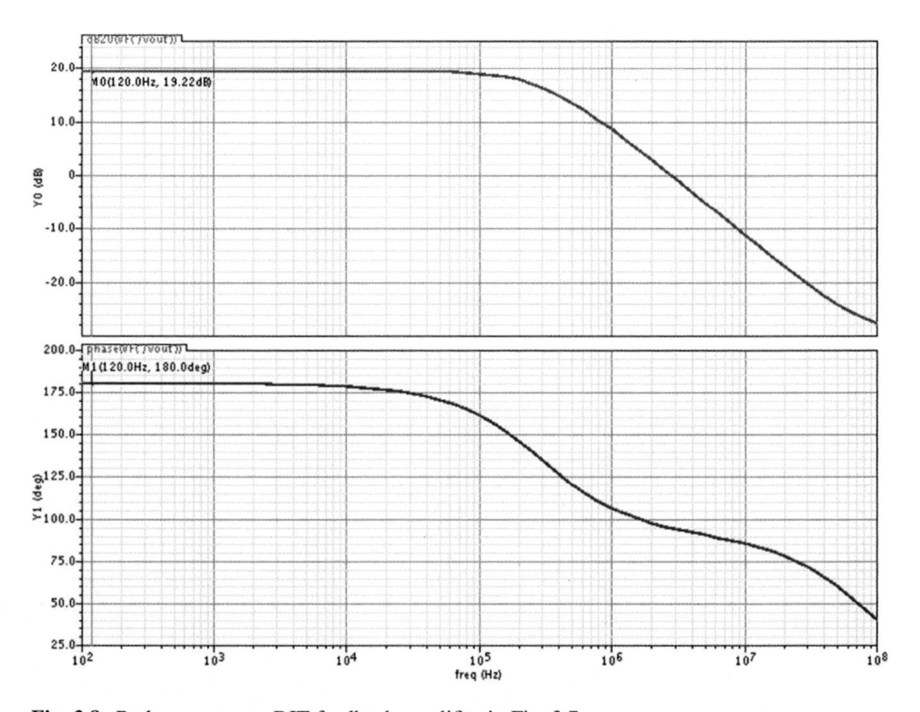

Fig. 3.8 Bode response to BJT feedback amplifier in Fig. 3.7

From this exercise, we see that calculating the transfer function for the single transistor feedback amplifier is no more difficult than finding one for a two-loop passive filter. Using two-port models for general amplifying cells results in the same characterization parameter set as for the bipolar (MOS, vacuum tube, etc.) and would be handled as well. We discuss this later on in this chapter.

Additional Analysis from the System Graph

The flow graphs we have generated contain more information than just the transfer function to the output node, node *2* in the examples above. Once generated, additional circuit responses are available from the same graph, and additional responses become available by "extending" the graph. We have already seen that transfer functions to both circuit nodes are obtained from the system flow. Additional results will be extracted in the next section. A word of caution is necessary at this point. While the relations in Eq. (3.4) are in the feedback relation "form" and the graph solution contains a "graph loop product" factor in the denominator, this factor is not necessarily the classic feedback loop gain function. It is simply algebra following the particular nodal selection and resulting DPI/SFG relations and has led to confusion in the literature [1–4]. This distinction will be fully developed in Chap. 5.

Branch Current and Power

Since we have the driving source voltage and both node voltages as graph arrows, we can generate the voltage response function across any branch by combining voltage arrows with a subtracting block to create the voltage across the branch, V_{12}, for example. To find a branch current between nodes *1* and *2*, we follow the subtracting block with the branch admittance block. These operations are shown in Fig. 3.9 where we have modified the graph to obtain the voltage across nodes 1 and 2 and then the current through y'_{12}. For the single transistor amplifier in

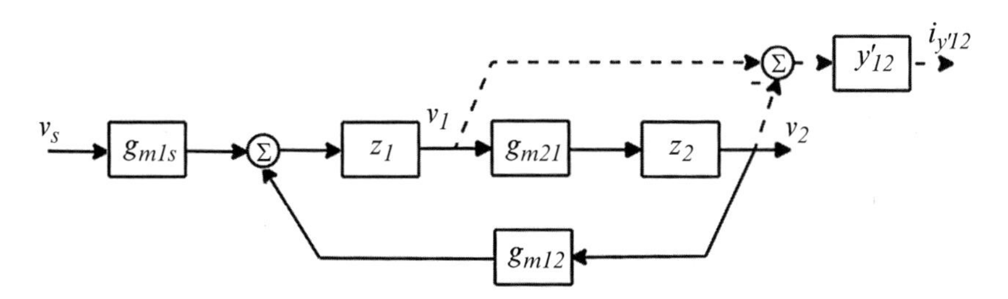

Fig. 3.9 Augmented flow graph to find the branch current in r_3

Fig. 3.7, the admittance y'_{12} can be the capacitance only, c_s; the resistor, g_f; or the sum, $(c_s + g_f)$, for the total branch current.

We can generate the power in y'_{12} by adding a multiplying block with inputs v_{12} and $i_{y'12}$. In a similar manner, we obtain the currents and power in every element or device.

Port Impedances

The impedance at a port "*x*" is found by setting independent sources to their internal impedance and either injecting a test current at node "*x*" and finding the resultant voltage or forcing a test voltage and finding the input current. We then take the ratio v_x/i_{test} to find the port impedance. Proceeding with our two-internal-node prototype topology, we modify the universal notation graph for these conditions as shown in Fig. 3.10 where we apply a test current to the short-circuit current summer at node 1. With this drive source, we solve for the node voltage v_1 to obtain the port impedance at node *1*.

Solving for v_1 and dividing by i_{test} yields the port impedance z'_1, where we indicate the port impedance as *z-prime* to distinguish it from the local DPI used in the graph, Eq. (3.12).

$$\frac{v_1}{i_{\text{test}}} = z'_1 = \frac{z_1}{1 - z_1 g_{\text{m21}} z_2 g_{\text{m12}}} \tag{3.12}$$

The port impedance at the base of the bipolar transistor for the circuit in Fig. 3.7 is found by substituting for the variables in this template solution with those from the amplifier circuit to obtain Eq. (3.13).

$$z'_1 = \frac{\dfrac{1}{g_{\text{s}} + g_{\text{f}} + cs}}{1 - \dfrac{(g_{\text{f}} + cs - g_{\text{m}})g_{\text{f}}}{(g_{\text{s}} + g_{\text{f}} + cs)(g_{\text{l}} + g_{\text{f}} + cs + h_{\text{oe}})}} \tag{3.13}$$

To find the port impedance at node 2, the collector node for this single transistor amplifier, we place the test current at node 2, adding a summer in the flow graph

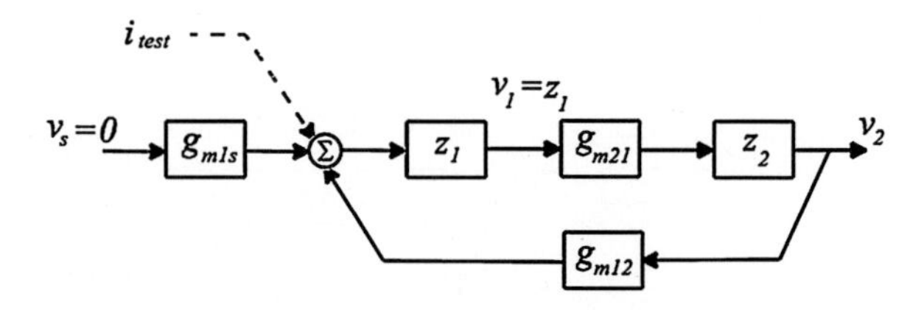

Fig. 3.10 Universal 2-node flow graph modified to find the port impedance at node 1

between the g_{m21} and z_2 blocks. We solve for v_2 and form the ratio defining z_2' template Eq. (3.14).

$$z_2' = \frac{z_2}{1 - z_1 g_{\mathrm{m21}} z_2 g_{\mathrm{m12}}} \tag{3.14}$$

Replacing the general parameters with the circuit-specific ones for the single bipolar transistor amplifier, we get Eq. (3.15) for the amplifier port impedance at the output node.

$$z_2' = \frac{\frac{1}{g_1 + g_f + cs + h_{oe}}}{1 - \frac{(g_f + cs - g_m) g_f}{(g_s + g_f + cs)(g_1 + g_f + cs + h_{oe})}} \tag{3.15}$$

Substituting parameters from the two-loop passive filter in Fig. 3.5 into the port impedance template for node *2*, Eq. (3.14), we obtain the port impedance for that circuit as given in Eq. (3.16).

$$z_{2,2\mathrm{loop\_filt}}' = \frac{\frac{1}{g_2 + \frac{1}{ls}}}{1 - \frac{\left(\frac{1}{ls}\right)^2}{\left(g1 + sc + \frac{1}{ls}\right)\left(g_2 + \frac{1}{ls}\right)}} \tag{3.16}$$

Relations in Eqs. (3.12) and (3.14) are seen to be general port *1* and port *2* impedances for feedback circuits with two non-driven nodes.

Extending the Flow Graphs to Include Embedded Cells

We are now ready to introduce a major extension to the DPI/SFG methodology: to include cells. An amplifier's equivalent y-parameter model fits directly into the short-circuit current/driving point impedance flow graph process. Consider the op amp inverting gain configuration shown in Fig. 3.11a and the equivalent small

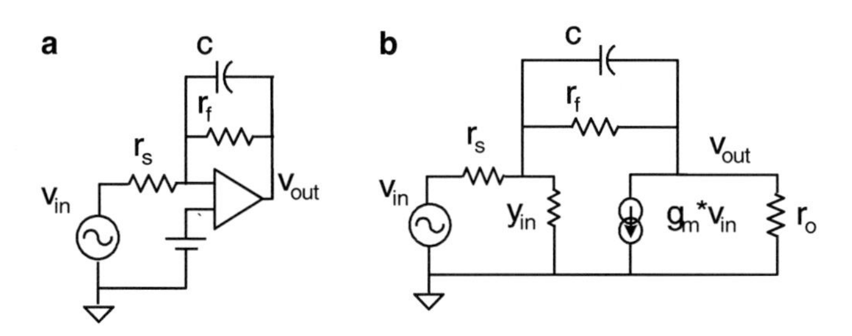

Fig. 3.11 (**a**) Inverting op amp gain cell and (**b**) y-parameter model for op amp

signal op amp and circuit model in Fig. 3.11b. Substituting this model for the amplifier allows us to find the flow graph parameters as Equation set (3.17):

$$g_{1s} = g_s; z_1 = \frac{1}{g_s + cs + g_f}; z_2 = \frac{1}{g_0 + cs + g_f}$$
$$g_{m21} = -g_m + cs + g_f; g_{m12} = cs + g_f \tag{3.17}$$

Substituting from Equation set (3.17) into the transfer function form, Eq. (3.4) reprinted here as Eq. (3.18) for easy reference, we quickly find the output transfer function for the inverting op amp gain stage as Eq. (3.19):

$$H_2(s) = \frac{g_{1s} \cdot z_1 \cdot g_{21} \cdot z_2}{1 - z_1 \cdot g_{21} \cdot z_2 \cdot g_{12}} \tag{3.18}$$

$$H_2(s) = \frac{\dfrac{g_s \cdot (g_f - g_m)}{(g_s + y_{in} + g_f + cs) \cdot (g_f + cs + g_o)}}{1 - \dfrac{(g_f + cs) \cdot (g_f - g_m)}{(g_s + y_{in} + g_f + cs) \cdot (g_f + cs + g_o)}} \tag{3.19}$$

For a “high-gain” op amp, the term in the denominator subtracted from one is large by design, allowing us to ignore the one in comparison. Most of the factors in the remaining expression cancel with those in the numerator with the final result becoming the well-known gain relation in Eq. (3.20):

$$H_2(s) = \frac{-g_s}{g_f + cs} \sim \frac{-r_f}{r_s} \cdot \frac{1}{1 + r_f cs} \tag{3.20}$$

The output impedance is found as Eq. (3.21). Input impedance is easily found using Eq. (3.9), substituting template variables with those specific to the circuit.

$$z_2'(s) = \frac{\dfrac{1}{(g_s + y_{in} + g_f + cs) \cdot (g_f + cs + g_o)}}{1 - \dfrac{(g_f + cs) \cdot (g_f - g_m)}{(g_s + y_{in} + g_f + cs) \cdot (g_f + cs + g_o)}} \tag{3.21}$$

The Single Transistor Amp: A Progression

The DPI/SFG process is now well described. There remains only learning how to implement it well. This is what we will do in the rest of the text. For the rest of this chapter, we will examine the single transistor amplifier using a sequence of progressively more complex models for the active element and adding parasitic capacitance to circuit nodes.

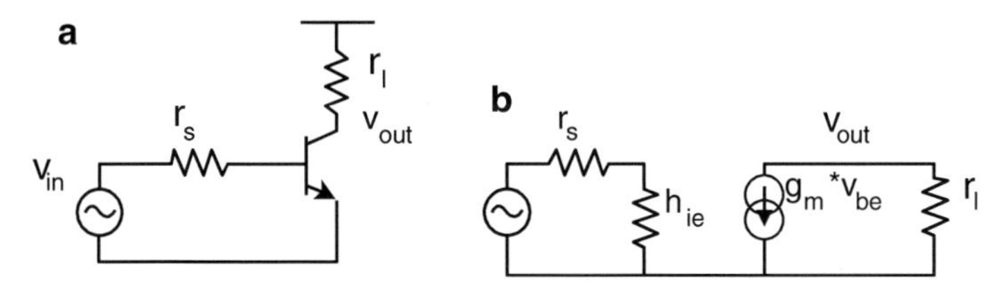

Fig. 3.12 (**a**) Single transistor amp and (**b**) small signal equivalent circuit

The Zero-Order Transistor Model

Consider the circuit in Fig. 3.12a along with its small signal equivalent circuit using the simple low-frequency transistor model, Fig. 3.12b. While the circuit is simple enough that we can write the transfer function by inspection, we resist the temptation creating instead the flow graph by inspection. There is no feedback in the circuit or in the simplified transistor model, so there will be no loop in the graph. Instead we find alternating transconductance–impedance blocks from the input node to the output node.

The transfer function is the product of the blocks from input to output given in Eq. (3.22).

$$\frac{v_{\mathrm{out}}(s)}{v_{\mathrm{in}}(s)} = \frac{g_{\mathrm{s}}}{g_{\mathrm{f}} + y_{\mathrm{ie}}} \cdot (-g_{\mathrm{m}} r_{\mathrm{l}}) \tag{3.22}$$

Converting Eq. (3.22) to "resistive" values instead of conductance, we get Eq. (3.23).

$$\frac{v_{\mathrm{out}}(s)}{v_{\mathrm{in}}(s)} = \frac{h_{\mathrm{ie}}}{r_{\mathrm{s}} + h_{\mathrm{ie}}} \cdot (-g_{\mathrm{m}} r_{\mathrm{l}}) \tag{3.23}$$

This last result in factored form reveals useful design information. The first factor is a voltage division at the input loop. To maximize the voltage gain, we maximize this factor by making $R_s < h_{ie}$. Similarly a large transconductance and large load impedance, the second factor, increases the gain.

Adding Elements to the Transistor Model and Nodal Capacitance

We improve the circuit modeling by improving the device model used by adding parasitic elements due to the physical construction. A good transistor will have a small output conductance, h_{oe}. If we include this device parameter in the model, the last block in the flow graph in Fig. 3.13 becomes $1/(h_{oe} + g_l)$, the parallel combination of

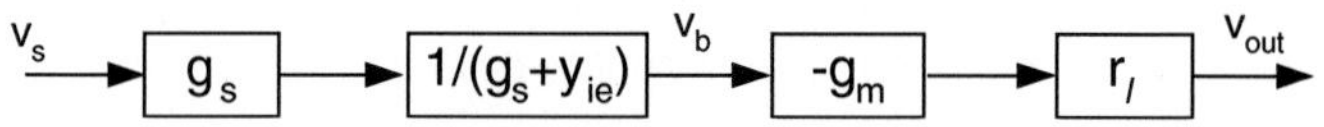

Fig. 3.13 Signal flow graph for circuit in Fig. 3.12

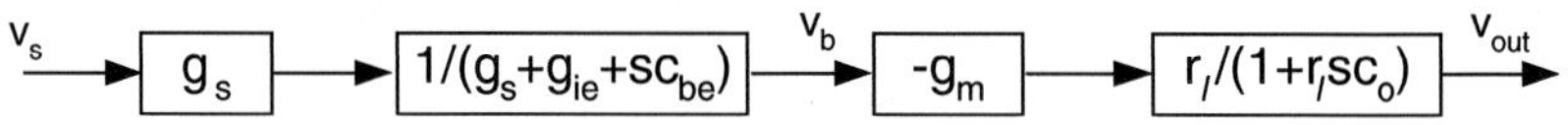

Fig. 3.14 Signal flow graph for circuit in Fig. 3.12

the output load r_l and the transistor's output resistance, $1/h_{oe}$, which will generally be dominated by $g_l(=1/r_l)$ so that we can usually ignore the effect of h_{oe}.

There will be parasitic capacitances at the base and collector nodes to AC ground. To obtain voltage gain, the output impedance will be larger than the input impedance so that a parasitic capacitance here will likely introduce the lower-frequency circuit pole as we increase the circuit's higher-frequency representation, producing the first deviation from the low-frequency analysis. Adding a capacitance from the output node to ground modifies the last block in the graph in Fig. 3.13 from r_l to $1/(g_l+sc_o)$. The transfer function now becomes as shown in Eq. (3.24),

$$\frac{v_{\text{out}}(s)}{v_{\text{in}}(s)}=\frac{h_{\text{ie}}}{r_{\text{s}}+h_{\text{ie}}}\cdot\left(-g_{\text{m}}\frac{r_{\text{l}}}{1+r_{\text{l}}sc_{\text{O}}}\right) \tag{3.24}$$

a single pole response at $\omega_0=1/(c_o r_l)$ due to the output node.

At higher frequencies, other poles affect the response. The transistor will have a base capacitance c_{be} that we add to the model to better emulate the transistor behavior at higher frequencies. Additionally, parasitic capacitance at the base node can be lumped in as an effective c_{be}, a net capacitance to ground. This modifies flow graph blocks as shown in Fig. 3.14.

The base node capacitance adds a second pole at $\omega_b=1/((r_s \| h_{ie})c_{be})$.

Adding Feedback to the Transistor Model

At high frequencies, the collector and base nodes couple through a junction capacitance, adding a feedforward/feedback path to the circuit. This effect is modeled by adding c_{bc} to the small signal transistor model in Fig. 3.12b as shown in Fig. 3.15. With this addition, the graph develops a loop characteristic of feedback circuits and systems such as we have already seen as in Fig. 3.4. We saw that this graph was applicable to the single transistor circuit with resistive feedback in Chap. 2. The present circuit, Fig. 3.12, with the inclusion of c_{bc} is topologically the same as the resistive feedback single transistor amplifier with c_{bc} replacing r_f

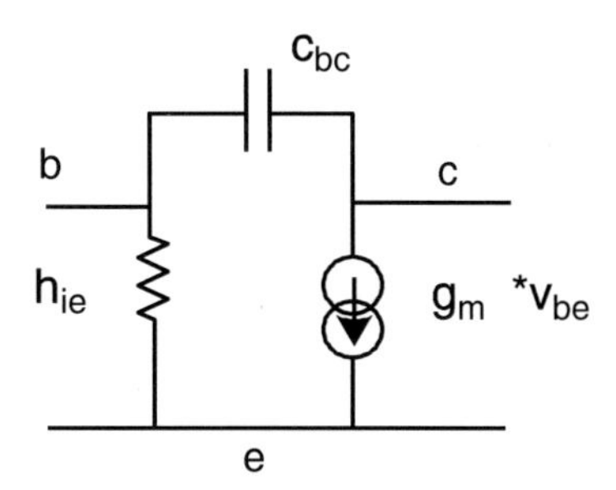

Fig. 3.15 Adding C_{bc} to the BJT small signal model

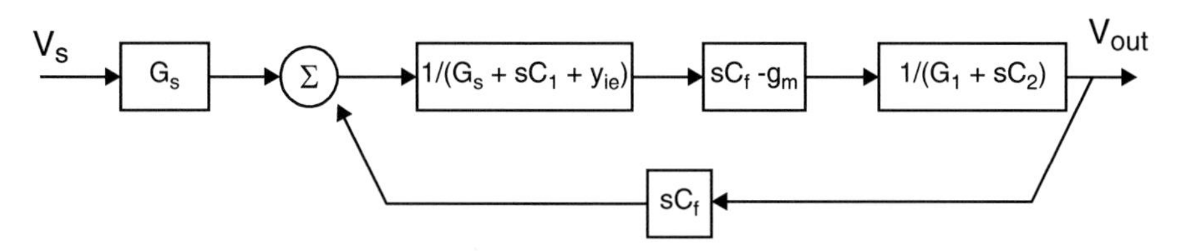

Fig. 3.16 Signal flow graph for circuit in Fig. 3.10 with capacitive feedback in model

(Fig. 3.16). Knowing this we find the circuit responses by replacing g_f with sc_{bc} in the results for the resistive feedback case. To reinforce the process, we develop the particular flow graph for the circuit in Fig. 3.12 with the additions of nodal capacitances and BJT base to collector capacitance.

The defining relations for the flow graph parameters for the two-internal-node circuit become as shown in Eq. (3.25),

$$
\begin{aligned}
&\mathrm{G}_{1s} = G_s \\
&Z_1 = \frac{1}{G_s + g_{ie} + sC_1}; Z_2 = \frac{1}{G_L + g_{oe} + sC_2} \\
&G_{21} = -g_m + sc_{bc}; G_{12} = sc_{bc}
\end{aligned} \tag{3.25}
$$

where we combine nodal capacitances as given in Eq. (3.26).

$$
C_1 = c_{bc} + c_{be}; C_2 = c_{bc} + c_o \tag{3.26}
$$

Solving this flow graph for the output transfer function, we get Eq. (3.27).

$$
\frac{v_{out}}{v_s} = \frac{G_s \cdot \frac{1}{G_s + g_{ie} + sC_1} \cdot (-g_m + sc_{bc}) \cdot \frac{1}{G_L + g_{oe} + s \cdot C_2}}{1 - (-g_m + sc_{bc}) \cdot \frac{1}{G_L + g_{oe} + s \cdot C_2} \cdot \frac{1}{G_s + g_{ie} + sC_1} \cdot sc_{bc}} \tag{3.27}
$$

Equation (3.27) shows that in addition to adding feedback and feedforward, c_{bc} adds to the nodal capacitances at the base and collector nodes, moving the poles at these nodes to lower frequencies.

The nodal port impedances become as shown in Eq. (3.28).

$$
\begin{aligned}
Z_2 &= \frac{\dfrac{1}{G_\mathrm{L} + g_\mathrm{oe} + s \cdot C_2}}{1 - (-g_\mathrm{m} + sc_\mathrm{bc}) \cdot \dfrac{1}{G_\mathrm{L} + g_\mathrm{oe} + s \cdot C_2} \cdot \dfrac{1}{G_\mathrm{s} + g_\mathrm{ie} + sC_1} \cdot sc_\mathrm{bc}} \\
Z_1 &= \frac{\dfrac{1}{G_\mathrm{s} + g_\mathrm{ie} + sC_1}}{1 - (-g_\mathrm{m} + sc_\mathrm{bc}) \cdot \dfrac{1}{G_\mathrm{L} + g_\mathrm{oe} + s \cdot C_2} \cdot \dfrac{1}{G_\mathrm{s} + g_\mathrm{ie} + sC_1} \cdot sc_\mathrm{bc}}
\end{aligned}
\tag{3.28}
$$

Z_I is the impedance for the circuit looking into the base of the circuit including the source resistance. We can remove the source impedance to find the port impedance as seen from the source looking into the base, set $G_s = 0$. Z_2 is the impedance looking into the output node.

Summary

In this chapter we used the DPI/SFG methodology to analyze a series of circuits that are topologically related and map onto the same flow graph form. It is due to this similarity that we are able to produce template graphs and template algebraic results, making it easy to modify a design and easily evaluate the effects of the changes. We extended the analysis to obtain additional circuit responses by adding algebraic branches to the graph, functions of nodal variables to obtain branch currents, and port impedances. We further extended the analysis to include cells such as amplifiers and treated these as easily as individual active elements by using their y-parameter representations. A particular nicety of the results obtained is the factored form showing correlation of circuit nodes to particular responses. Output/input low-frequency gains and frequency responses were obtained. Poles were found to be due to individual node impedances and could be defined by design to a certain extent. This fact will be shown to be very useful in compensation of amplifiers to insure stability in later chapters.

We will return to these topics and examples in later chapters as we further develop these properties of analysis using the DPI/SFG approach. Note the ease in which feedback in these circuits has been handled as a natural application of the method without the need of even recognizing that the circuits had feedback at all. We did not need to invoke the ideal parameters of "open-loop gain" and feedback factors in the discussions. These parameters will be discussed and extracted in the later chapters particularly in the discussions on circuit stability.

References

1. P.R. Gray, R. Meyer, *Analysis and Design of Analog Integrated Circuits* (John Wiley and Sons, New York, NY, 1977)
2. A. Ochoa, *Analyzing Feedback: Properly Simulating the Open Loop. Mid West Symposium on Circuits and Systems* (IEEE, Notre Dame, IN, 1998)
3. R.D. Middlebrook, Measurement of loop gain in feedback systems. Int. J. Electron. **38**(4), 485–512 (1975)
4. A. Ochoa, *Loop Gain in Feedback Circuits: A Unified Theory Using Driving Point Impedance. Mid West Symposium on Circuits and Systems* (IEEE, Notre Dame, IN, 2013)

Chapter 4
Noise and Mismatch

Abstract Noise and device mismatch are circuit properties that limit the useful range of a design and can be minimized by proper design. Both are statistical; one is dynamic and actively present in use (noise), while the other is stable differing from one specific chip to another and introducing offsets in the circuit response. In order to monitor the impact on the circuit of these quantities, we translate the equivalent signal from its place of introduction to either the input or the output where it is compared with the signal. How we do this translation and how we combine signals from different sources is the subject of this chapter.

Noise and mismatch are seemingly quite different circuit effects—noise is dynamic appearing as a time-varying corruption to the signal processing impacting dynamic range, while mismatch represents stable shifts in DC operating conditions appearing as offsets impacting yield. Both are statistical in nature however and are normally reflected onto the input of a circuit to gage its impact in the design. As such they can be operated on in the same way using DPI/SFG. In this chapter, we will show how to "move" a signal from its introduction in the circuit flow graph to input or output for quantitative comparison with alternative designs to meet a specification.

Noise Signals in Circuits

Noise is dynamic and statistical in nature—it appears as random fluctuations at a device terminal as an open-circuit voltage or as a short-circuit current. A time plot of a noise signal would appear as shown in Fig. 4.1.

The signal is random with its magnitude jumping around with no correlation to previous time samples and, having as much positive contributors as negative, averaging over time to zero. A subsequent time capture will produce another such trace, also not correlated to any other time plot. These properties are expressed mathematically as Eq. (4.1), zero mean, and as Eq. (4.2), zero dot product or uncorrelated between samples or with the same sample shifted in time.

A. Ochoa, *Feedback in Analog Circuits*, DOI 10.1007/978-3-319-26252-9_4

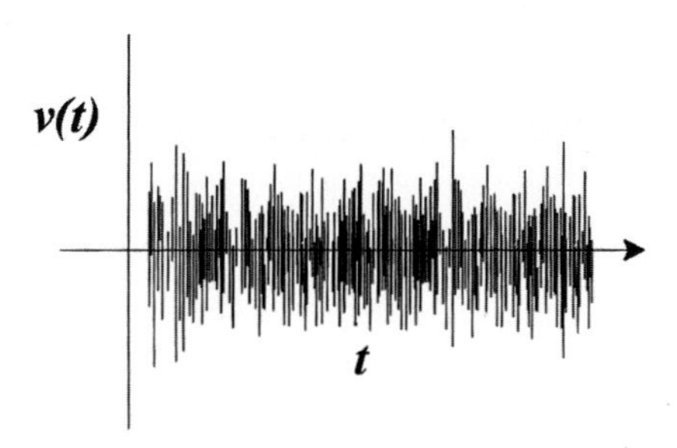

Fig. 4.1 A noise signal versus time

$$< x >= \frac{1}{T}\int_T x(t)\mathrm{d}t = 0 \tag{4.1}$$

$$< x_1(t) \cdot x_2(t) >= \int_T x_1(t) \cdot x_2(t)\mathrm{d}t = 0 \tag{4.2}$$

With these odd characteristics, how do we quantify noise effects in circuits? Since the signals are not repeating, we cannot insert a time waveform into our simulator and process the response. From a family of noise traces, we formulate a statistical measure of the noise and process this statistical measure. Having zero mean, averages are not useful as the average of a family of traces having zero mean results in zero as well. The average of the square of a signal current or voltage is proportional to power in circuit analysis, and in statistics such operations generate the variance of the variable. If then we take a family of voltage noise plots, square each time value, and then average the whole set, we obtain a signal proportional to the effective power delivered to the circuit having a statistical significance. Since power is the product of current and voltage, we need to transform one of the voltage signals to a current. We do this in a normalized fashion by dividing by 1 Ω (Eq. 4.3).

$$< p_{1\Omega} >=< \frac{1}{T}\int_T \frac{v^2(t)}{1\Omega}\mathrm{d}t > \tag{4.3}$$

Equation (4.3) represents the average of many time waveforms of normalized power measurements producing the average variance of the noise voltage (current). The square root of this integral yields the noise voltage (current) spread or noise voltage average standard deviation (sigma). In practice, the result in Eq. (4.3) is obtained using a frequency space measurement, a power spectrum analyzer, and the identity between time domain and frequency domain of power calculation stated in Eq. (4.4).

$$\frac{1}{T}\int_T \frac{v^2(t)}{1\Omega}\mathrm{d}t = \int_f \mathrm{PSD}_{1\Omega}(f)\mathrm{d}f \tag{4.4}$$

In the frequency domain, specified as a power density, PSD, we integrate over the bandwidth of interest to obtain the effective averaged 1 Ω voltage variance. A current PSD similarly results in the ensemble time average current squared or variance of the current average. Next we explore how to use these results.

Noise in Simple RC Filter

Resistors have a noise PSD that is proportional to temperature given as Eq. (4.5) where "k" is the Boltzmann constant (1.38×10^{-23} J/K), T is the temperature in degrees Kelvin, "r" the resistance magnitude, and Δf the differential bandwidth (1 Hz). For the current PSD, the resistance is replaced with the conductance g, Eq. (4.5).

$$\begin{aligned} v_{\mathrm{r}}^2(s) &= 4kTr\Delta f \\ i_{\mathrm{r}}^2(s) &= 4kTg\Delta f \end{aligned} \tag{4.5}$$

The voltage and current PSDs are constant with frequency as depicted in Fig. 4.2 for the equivalent noise voltage and noise current.

A 50 Ω resistor at 300 °K will have 0.91 nV rms noise voltage or 0.0182 nA rms noise current.

Since reactive elements do not dissipate energy, they do not introduce noise power into the circuit. For the simple RC filter in Fig. 4.3a, we represent the noise as its current equivalent resistor noise and the system signal flow graph as shown in Fig. 4.3b.

The noise current source is a floating source with one end attached to a low-impedance signal source. That side does not contribute to the short-circuit current, i_{sc}, at the output node. The output voltage signal is found using the flow graph as Eq. (4.6) where the circuit shapes the noise input signal from its flat, wideband form to that of a single-pole response as shown in Eq. (4.6).

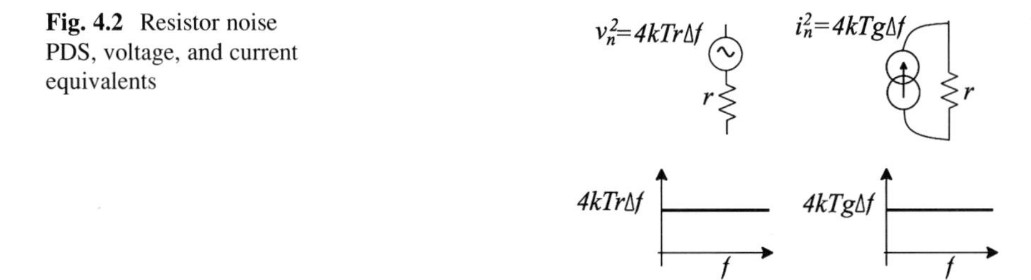

Fig. 4.2 Resistor noise PDS, voltage, and current equivalents

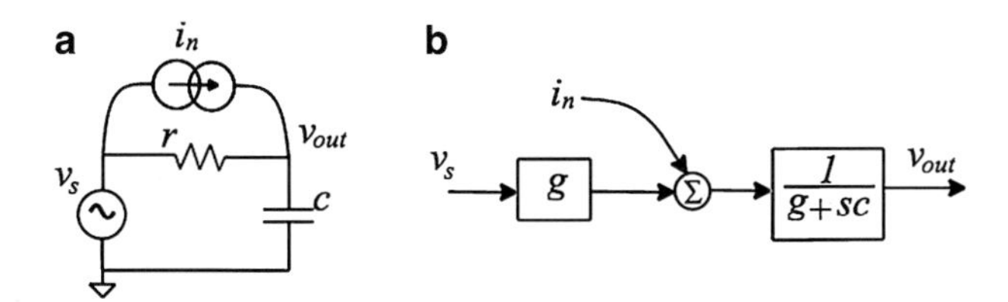

Fig. 4.3 (**a**) RC filter with current noise source, (**b**) system flow graph

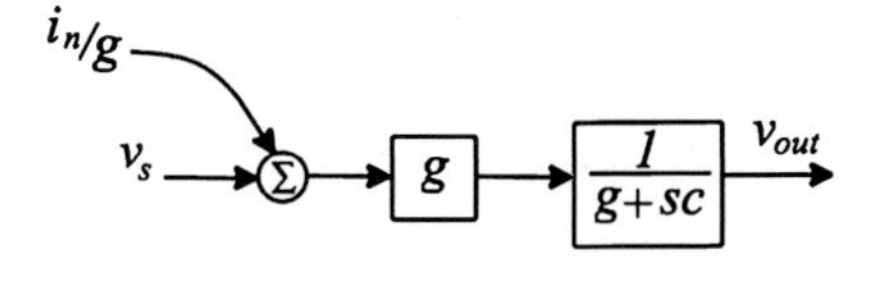

Fig. 4.4 Signal flow graph for RC filter after moving noise source to the input

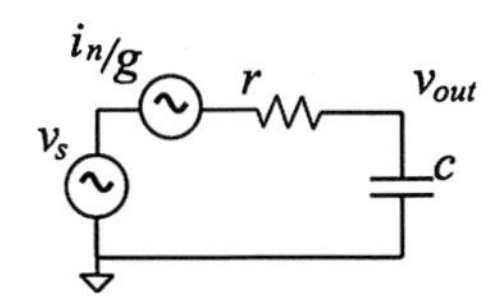

Fig. 4.5 RC filter with noise signal moved to input

$$\begin{aligned} v_{\text{out}}(s) &= i_{\text{n}} \cdot \frac{1}{g_{\text{s}} + c \cdot s} \\ &= \sqrt{4kTr_{\text{s}}(1\,\text{Hz})} \cdot \frac{1}{1 + r_{\text{s}}c \cdot s} \end{aligned} \tag{4.6}$$

In order to compare a noise signal with the input signal, we want to find an equivalent noise source at the input such that this signal produces the same output noise. The flow graph represents algebra, and as such we can perform flow graph manipulations maintaining input/output integrity. Basic flow graph algebra requires that we multiply a signal by the contents of a graph box if we move in the direction of the arrow. Moving against the arrow we divide the signal by the box contents. We can "move" the noise current signal backward by modifying the graph in Fig. 4.3b as shown in Fig. 4.4.

This flow graph represents the circuit in Fig. 4.5.

We now have the noise signal "referred to the input" with the noise removed from the circuit itself—now considered noiseless. For circuits having multiple noisy elements, we treat each noise source separately and can then translate each to the input (or output) where they can be combined into one effective source. Since noise is statistical, we combine these components as statistical variables finding the net variance as the sum of the translated PSDs, the squares of the noise signals. Noise from independent sources is uncorrelated so we add the PSDs and take the square root of the sum to find the equivalent voltage or current signal. Correlated noises are algebraically combined before doing the RSS operation with other signals.

Having the noise signal at the input however is not sufficient. The input resistor noise is wideband, while the source signal will be band-limited or even a single frequency. How do we compare signal to noise then? From Eq. (4.4), the variance of the noise voltage is found by integrating the normalized PSD over frequency. At the input, the referred noise may see the different impedance of the source supply so that the frequency response here is different than it is at the output node. Also the noise at the input is not real—it is a mathematically derived quantity such that it will produce the same noise at the output node of the noiseless circuit as the real circuit does. Putting a spectrum analyzer at the input node will not result in a measurement of this input-referred noise, while that at the output is real—it can be calculated and measured as the variance of the noise voltage at the output node, (Eq. 4.7).

$$< v_{\text{out}}^2 >= \int_f 4kTr_s \cdot \left| \frac{1}{1 + r_s c \cdot j2\pi f} \right|^2 \mathrm{d}f \tag{4.7}$$

We see in Eq. (4.7) that the circuit band limits the noise so that the high frequencies are removed from impacting the circuit behavior. We "integrate" this output noise over frequency to convert the noise to an effective noise variance. Taking the square root of the variance results in a statistical value that the bulk of the noise amplitude will have a probability of ~68 % of being less than this value—the ±1 sigma value.

To integrate Eq. (4.7) over frequency, we use the identity in Eq. (4.8) to obtain Eq. (4.9).

$$\int \frac{1}{1 + x^2} \mathrm{d}x = \tan^{-1}(x) \tag{4.8}$$

$$< v_{\text{out}}^2 \geq \frac{kT}{c} \tag{4.9}$$

This result states that the 1-sigma $<v_{out}>$ deviation due to noise at the output of the *r-c* filter is only a function of capacitance *c* and not of the resistance that generates the noise. For a given capacitance, increasing the resistance generates more noise per unity bandwidth as it reduces the circuit bandwidth proportionally keeping the integral constant.

The system flow graph is now as shown in Fig. 4.6.

By performing the integration over frequency, we have converted the noise signal from a frequency-dependent one to the statistical variance. The square root of the variance is the 1-sigma noise voltage magnitude superimposed on the desired output signal (Fig. 4.7).

With the noise signal transformed to a variation, moving the signal back through the flow graph as in Fig. 4.4 is not strictly valid. We are moving a statistical constant through the graph that adds to the signal at every point making the effective signal ~78 % contained by $v_s \pm v_n$; noise rides on top of the system signal producing an

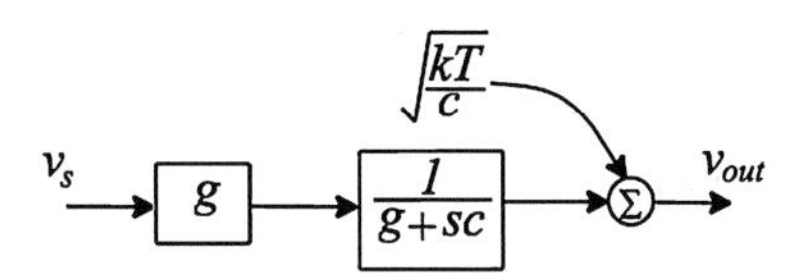

Fig. 4.6 Flow graph for RC filter with noise at output

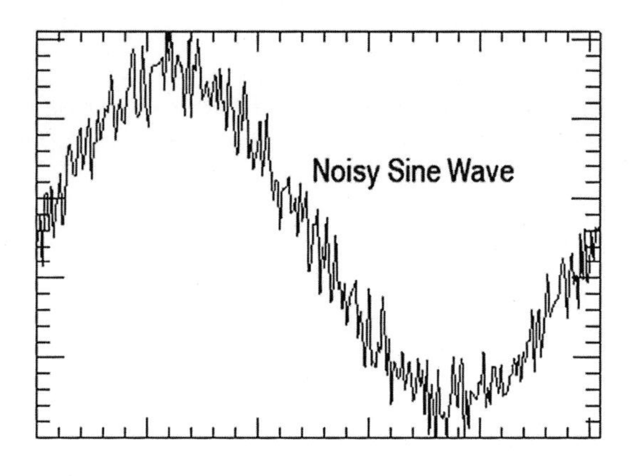

Fig. 4.7 Noisy sine wave

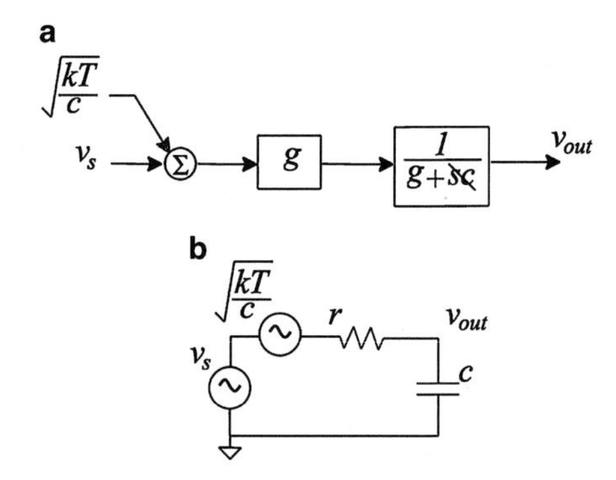

Fig. 4.8 (**a**) Flow graph with input-referred total noise and (**b**) RC filter with input-referred noise

instantaneous error or deviation from the signal with statistical standard deviation v_n. We modify the flow graph by stipulating the frequency (Laplace variable s) to be that of the signal of interest. For the low-pass filter, any frequency in the signal band can be used. A convenient frequency is 0 Hz. Setting the Laplace variable s to zero, we can move the noise output response to the input by multiplying the signal first by g then by "$1/g$" as we cross each of the blocks, resulting in the flow graph and circuit in Fig. 4.8 with input-referred 1-sigma equivalent noise voltage. For a filter with $10\ pF$ capacitance, the noise voltage signal is about $20\ \mu V$ rms.

We have seen that a circuit band limits the noise defining the effective variation over this bandwidth. Any signal within this frequency range will have noise perturbations of that magnitude so we translate the output noise to the input to any frequency in this band. For a low-pass design, using 0 Hz simplifies the process. For a bandpass, center frequency will be appropriate; for a high pass, any frequency in the interest band will do.

Using a circuit with a single noise source, the r-c filter, we have walked through the process of analyzing noise as:

1. Add a noise source (PSD) for each circuit element that generates noise.
2. Find the noise PSDs at the output node by common circuit analysis means. Combine the individual noise PSDs to form the net noise power at the output—correlated signals are first algebraically combined before squaring and combining with other signals that are not correlated to them.
3. Translate the effective output PSD back through the graph to the input to provide "spot" input-referred noise power. Noise signal is found by taking the square root of the PSD.

Alternatively, after finding the net effective PSD at the output, we can integrate the output noise squared over frequency to obtain the statistical representation for the effective noise signal at the output and then translate the effective noise variance to the input by moving the signal backward through the circuit's flow graph with frequency elements set to the signal frequency. This process provides the appropriate bandwidth for finding the effective noise variation, as the circuit bandwidth at the input is not necessarily the same as that at the output. We find the signal's rms value to compare with the input-referred effective noise variation.

Noise from Active Devices

Devices have noise properties defined by the physical composition of the device. We have seen that resistor's have a flat, wideband noise defined by physical constants, temperature, and its resistance (conductance) value. This noise is termed "thermal noise." Other noise types of device interest are shot and flicker noise.

Shot noise is a consequence of the random transition of charges across a potential barrier, modeled as having a time averaged variance as given in Eq. (4.10).

$$< i_{\mathrm{sh}}^2 >= 2qI\Delta f \tag{4.10}$$

I in Eq. (4.10) is the dc current, q the electronic charge, and Δf the measurement bandwidth. Shot noise PSD is flat with frequency.

Flicker noise shows a spectrum as given in Eq. (4.11)

$$< i_{\mathrm{f}}^2 >= \frac{K_{\mathrm{f}} I^m \Delta f}{f^n} \tag{4.11}$$

where I is the dc current, K_{f} is the flicker noise coefficient, m the flicker noise exponent, and n the frequency exponent, approximately 1. Flicker noise is associated with charge trapping and releasing at the Si-oxide interface and varies widely with technology. In MOS devices, flicker noise varies as $1/(WL)$ making the use of large transistors useful in reducing this component of noise. The PSD for flicker noise varies as the reciprocal of frequency ($n \sim 1$) and is also known as $1/f$ noise. An electronic device is modeled using combinations of these noise types depending on the device composition and operating physics.

Diode Noise

The diode has a combination of resistive noise due to the structure in series with the bipolar junction, and shot and 1/*f* noise for the junction itself (Fig. 4.9).

$<v_s^2>$ is a thermal noise due to bulk regions in series with the junction and $<i_d^2>$ is a combination of junction shot and flicker noise, Eqs. (4.12) and (4.13).

$$< v_{\mathrm{s}}^2 >= kTr_{\mathrm{s}}\Delta f \tag{4.12}$$

$$< i_{\mathrm{d}}^2 >= 2qI_{\mathrm{D}}\Delta f + \frac{KI_{\mathrm{D}}^{\alpha}\Delta f}{f} \tag{4.13}$$

$r_{\mathrm{d}} = kT/(qI_{\mathrm{d}})$ is the diode diffusion resistance and not a real resistor, while r_{s} the diode series resistance is due to physical silicon bulk volume and contacts.

Bipolar Transistor Noise Model

The bipolar transistor has contributions to noise due to bulk resistance, shot noise at energy barriers, and flicker noise due to fluctuations in carrier recombination in the space charge regions, modeled as shown in Fig. 4.10.

MOSFET Noise Model

MOSFET devices have both flicker and thermal noise contributors. The channel region provides the dominate thermal noise, while the semiconductor-oxide interface region generates flicker noise. We can add resistor thermal noise due

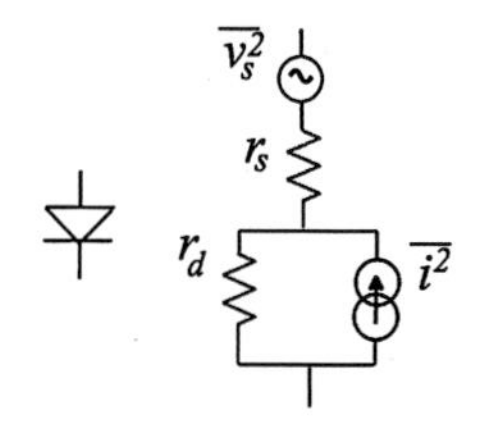

Fig. 4.9 Noise model for the junction diode

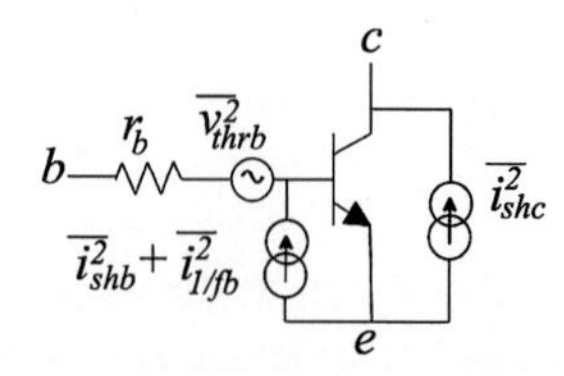

Fig. 4.10 Noise model for the bipolar transistor

to series source and drain resistances and shot noise due to oxide leakage current, but these are usually small and not included here for simplicity. A MOSFET can be modeled as shown in Fig. 4.11 with the effective drain noise current given in Eq. (4.14),

$$< i_{\mathrm{d}}^2 >= 4kT(\gamma g_{\mathrm{m}})\Delta f + \frac{K_{\mathrm{f}} I_{\mathrm{D}}^{\propto} \Delta f}{f^n} \tag{4.14}$$

where:

γ is the channel factor ~2/3.
g_{m} is the device transconductance.
K_{f} is the flicker coefficient.
α is the flicker drain current exponent ~1.
n is the flicker frequency exponent ~1.

Noise in a Single Transistor Amplifier

Figure 4.12 shows a single transistor amplifier with three noise sources due to the two resistors, load and source, and the nMOS transistor. The transistor noise is a composite noise i_d, combining channel thermal and flicker noise as in Eq. (4.14).

We first combine the noise sources at the output node, i_n Eq. (4.15), as the sum of squares since the noise signals are uncorrelated. The flow graph for the circuit in Fig. 4.12 is shown in Fig. 4.13 after combining the noise sources at the output node.

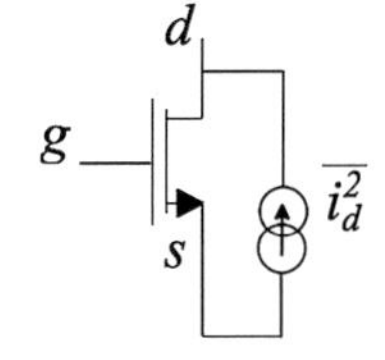

Fig. 4.11 Noise model for MOSFETs

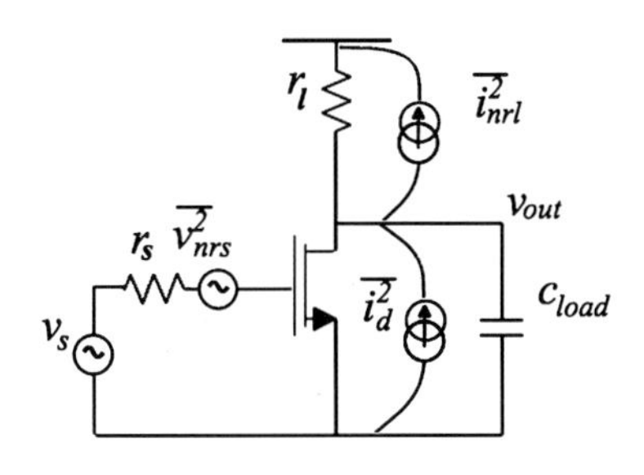

Fig. 4.12 Single transistor amp with single noise sources

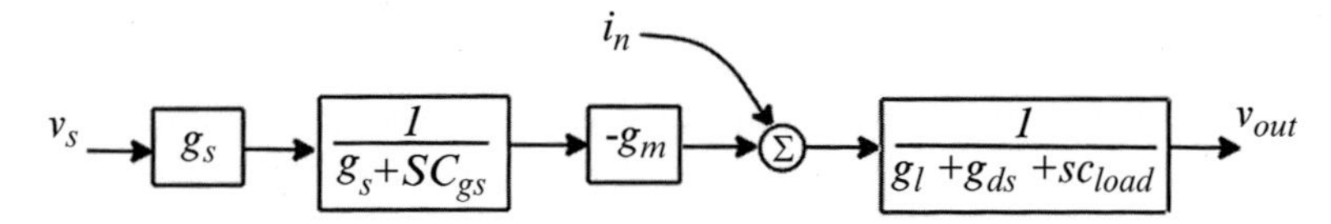

Fig. 4.13 Flow graph for single transistor amp of Fig. 4.13 (**a**)

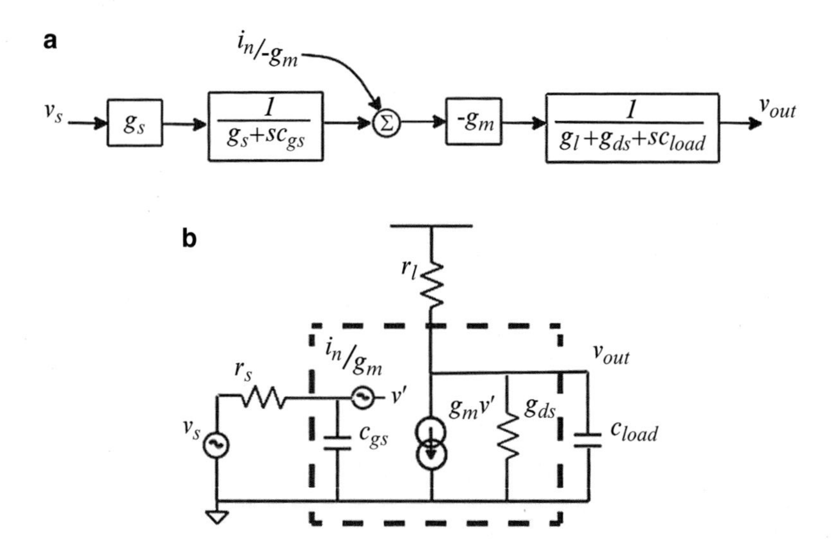

Fig. 4.14 (**a**) Modified flow graph in Fig. 4.14 with the noise current moved one block left and (**b**) modified equivalent circuit

$$i_{\mathrm{n}}^2 = i_{\mathrm{nrl}}^2 + i_{\mathrm{d}}^2 + i_{\mathrm{nrs}}^2 \tag{4.15}$$

Moving the i_n noise signal in the flow graph back toward the input through one block transforms the current signal to a voltage signal (Fig. 4.14a). We interpret this flow graph as the small-signal equivalent circuit shown in Fig. 4.14b. The modified noise source is added to the node voltage at the gate of the nMOS transistor making it a floating source *inside* the small-signal model [1] of the transistor where it draws no current.

Moving the noise signal by one block again toward the input converts the voltage representation back to a current, now in shunt at the transistor input (Fig. 4.15). This spot noise, i_{nio} in Eq. (4.16), is seen to increase at high frequencies. This increase is necessary in order to keep the modeled noise at the output the same as before referring it to the input. Remember this is a "mathematical" noise and not a real noise.

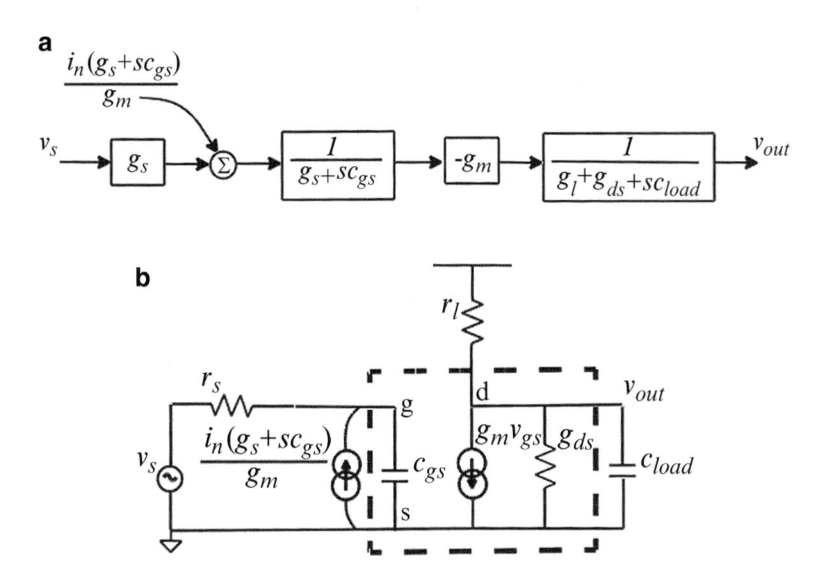

Fig. 4.15 (**a**) Modified flow graph in Fig. 4.14 with the noise current moved two blocks left and (**b**) modified equivalent circuit

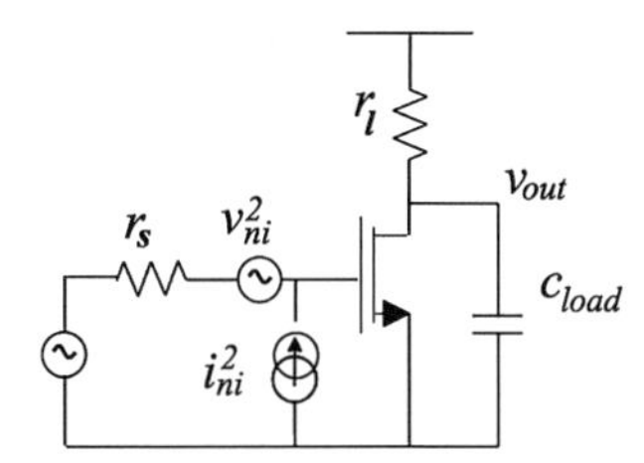

Fig. 4.16 One transistor amp with noise referred to input as the two-source equivalent

$$i_{\mathrm{ni0}}^2 = \frac{i_{\mathrm{n}}}{-g_{\mathrm{m}}} \cdot \left(g_{\mathrm{s}} + sc_{\mathrm{gs}}\right) \tag{4.16}$$

We have now transformed the noise currents from the circuit output to an equivalent noise current at the input. We can move this current again across the g_s block converting the shunt input current to a series voltage noise source in the input loop. In this form it is functionally at the same level as the circuit input signal v_s allowing for direct comparison with the input signal.

Equivalent noise at the input is usually given as two sources, one a voltage in series with the input signal source v_s and the other a current source in shunt with the input node as shown in Fig. 4.16. To put it into this dual-source form as found in the literature [2], we note that the shunt input noise current source (i_{nio}) consists of two terms and move only the term that is proportional to g_s backward across the g_s block making this part a voltage signal v_{ni}. The other term remains a shunt current i_{ni}. Algebraically we get Eq. (4.17).

$$\begin{aligned} i_{\mathrm{ni0}}(s) &= \frac{i_{\mathrm{n}}}{-g_{\mathrm{m}}} \cdot (g_{\mathrm{s}} + sc_{\mathrm{gs}}) \\ i_{\mathrm{ni}}(s) &= \frac{i_{\mathrm{n}}}{-g_{\mathrm{m}}} \cdot sc_{\mathrm{gs}} \\ v_{\mathrm{ni}}(s) &= \frac{i_{\mathrm{n}}}{-g_{\mathrm{m}}} \end{aligned} \tag{4.17}$$

The process has been direct and algebraically complete making the results self-consistent. v_{ni} is seen to be correlated with i_{ni}, both noise sources being proportional to the output current noise source i_n. The signs are relevant to properly combine the effects from these sources on system responses.

Noise Translations in a Feedback Circuit

Keeping with the single transistor amplifier, we add resistive feedback (Fig. 4.17). The added feedback resistor r_f contributes additional noise to the circuit i_{nrf}. We will use this example to show how noise signals are first moved to the output where they are statistically combined into the effective output noise voltage in a feedback design. This voltage will be integrated to provide the variance of the noise voltage that will then be moved to the input loop where it can be compared to the input signal power level.

The transistor noise is shown as $<i_d^2>$, the net transistor noise in Fig. 4.9 after combining the thermal and shot noises. The individual noise signals are added to the circuit flow graph in Fig. 4.18 still as current and voltage signals in order to move them individually through the graph to the output where they will be appropriately combined. Resistor noise can be represented as either a voltage source in series or as a current source in shunt with the resistor. The noise for resistor r_s is shown as a series voltage equivalent while noise from the feedback resistor appears as a shunt current in the schematic and as two currents in the flow graph, i_{nrf} and $-i_{nrf}$. The negative sign is needed to maintain proper correlation between these two signals.

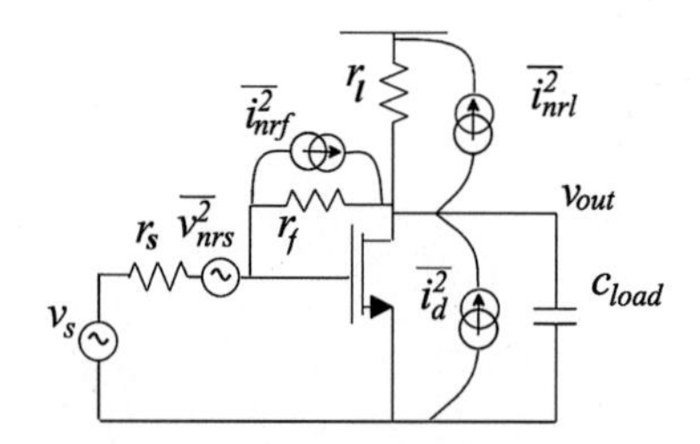

Fig. 4.17 Single transistor amplifier with feedback

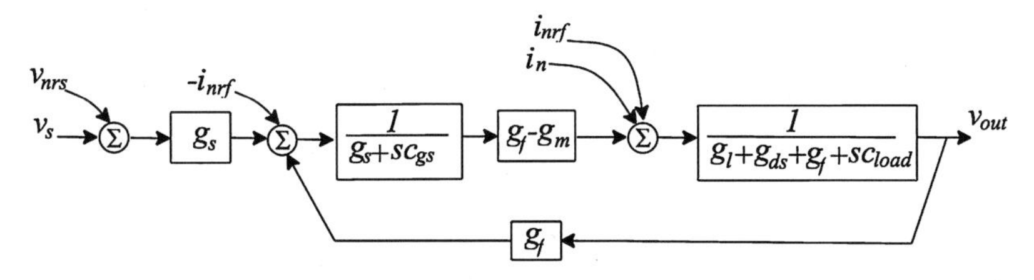

Fig. 4.18 Signal flow graph for feedback amplifier in Fig. 4.18 showing device noise signals as current and voltage signals

The signal i_n arises from combining transistor noise $<i_d^2>$ with the load resistor noise $<i_{nrl}^2>$.

A signal is moved from its flow graph point of entry to the flow graph output by multiplying it by the net transfer function it sees between these two points. We find the magnitude squared of this product (for uncorrelated noises) before combining with other noise powers at the output. Equation (4.18) shows the translated noise power at the output, v_{nrso}, for the source resistor noise v_{nrs},

$$\begin{aligned} < v_{\mathrm{nrso}}^2 >&= \left|(H_{\mathrm{onrs}})\right|^2 \cdot < v_{\mathrm{nrs}}^2 > \\ &= \left|\frac{\left(\frac{g_{\mathrm{s}}}{g_{\mathrm{s}}+g_{\mathrm{f}}+sc_{\mathrm{gs}}}\right)\left(\frac{g_{\mathrm{f}}-g_{\mathrm{m}}}{g_{\mathrm{f}}+g_{\mathrm{l}}+g_{\mathrm{ds}}+sc_{\mathrm{load}}}\right)}{1-\mathrm{LP}}\right|^2 \cdot < v_{\mathrm{nrs}}^2 > \end{aligned} \tag{4.18}$$

where LP is the loop product and given as given in Eq. (4.19).

$$\mathrm{LP} = g_{\mathrm{f}} \cdot \left(\frac{g_{\mathrm{s}}}{g_{\mathrm{s}}+g_{\mathrm{f}}+sc_{\mathrm{gs}}}\right)\left(\frac{g_{\mathrm{f}}-g_{\mathrm{m}}}{g_{\mathrm{f}}+g_{\mathrm{l}}+g_{\mathrm{ds}}+sc_{\mathrm{load}}}\right) \tag{4.19}$$

All the other noise signals translate similarly by finding the transfer function for that signal, squaring it, and multiplying by the noise PSD. Note that there are two input signals for the feedback resistor noise i_{nrf}. Here we transform each and combine them algebraically before squaring the net transfer function to maintain correlation

$$< i_{\mathrm{nrfo}}^2 >= \left|H_{\mathrm{rnfd}} - H_{\mathrm{rnfs}}\right|^2 \cdot 4kTg_{\mathrm{f}}\Delta f \tag{4.20}$$

where H_{rnfd} is the transfer function for noise signal at the drain side of r_f and H_{rnfs} for the signal at the source side (Eqs. 4.21 and 4.22).

$$H_{\mathrm{rnfd}} = \frac{\left(\frac{1}{g_{\mathrm{f}}+g_{\mathrm{l}}+g_{\mathrm{ds}}+sc_{\mathrm{l}}}\right)}{1-\mathrm{LP}} \tag{4.21}$$

$$H_{\mathrm{rnfs}} = \frac{\left(\dfrac{1}{g_{\mathrm{s}} + g_{\mathrm{f}} + sc_{\mathrm{gs}}}\right)\left(\dfrac{g_{\mathrm{f}} - g_{\mathrm{m}}}{g_{\mathrm{f}} + g_{\mathrm{l}} + g_{\mathrm{ds}} + sc_{\mathrm{l}}}\right)}{1 - \mathrm{LP}} \tag{4.22}$$

The net output *1* Ω noise power per unit frequency (spot) is then found as the sum of individual noise powers at the output Eq. (4.23).

$$\begin{aligned} < v_{\mathrm{nout}}^2 > &= \left|H_{\mathrm{id}}\right|^2 < i_{\mathrm{d}}^2 > + \left|H_{\mathrm{nrl}}\right|^2 < i_{\mathrm{nrl}}^2 > \\ &+ \left|H_{\mathrm{nrfd}} - H_{\mathrm{nrfs}}\right|^2 < i_{\mathrm{nrf}}^2 > + \left|H_{\mathrm{nrs}}\right|^2 \cdot < v_{\mathrm{nrs}}^2 > \end{aligned} \tag{4.23}$$

We now integrate the spot noise, Eq. (4.23), over frequency to obtain the variance of power into *1* Ω as we did with the simple r-c filter, Eq. (4.13), and take the square root, Eq. (4.24), to obtain the effective standard deviation for the noise voltage at the output node.

$$< v_{\mathrm{nout}} >= \sqrt{\int_0^{\infty} v_{\mathrm{nout}}^2(f)\mathrm{d}f} \tag{4.24}$$

We now have the noise voltage standard deviation magnitude at the output—a *1*-sigma error band around our noise-free system signal. The circuit will amplify the system signal so that direct comparison of this noise magnitude with the signal will depend on the gain of circuit. To remove this dependence, we move the noise magnitude to the input where a direct comparison of the square root of the signal power can be done. We divide the net output noise voltage (sigma) by the gain of the amplifier in its pass band, here DC, Eq. (4.25).

$$\begin{aligned} Gain - dc &\sim -\frac{r_{\mathrm{f}}}{r_{\mathrm{s}}} \\ < v_{\mathrm{ns}} >&= \frac{< v_{\mathrm{nout}} >}{\left|-\dfrac{r_{\mathrm{f}}}{r_{\mathrm{s}}}\right|} \end{aligned} \tag{4.25}$$

We have found the noise voltage standard deviation at the output and referred it to the input by dividing by the stage pass-band gain (Eq. 4.25). It is a single voltage source, Fig. 4.19, not the dual source as shown in Fig. 4.16. We do not need to transform this result to the two-source equivalent input noise source combination as it is complete in itself. To find the two-source equivalent and show how it can be done, we first solve for the noise v_{nin} at the output of the amplifier input loop due to v_{ns} using the graph in Fig. 4.20.

The equivalent noise voltage at the source v_{ns} in Fig. 4.20a is moved to the output of the amplifier input loop as v_{nin} given as in Eq. (4.26).

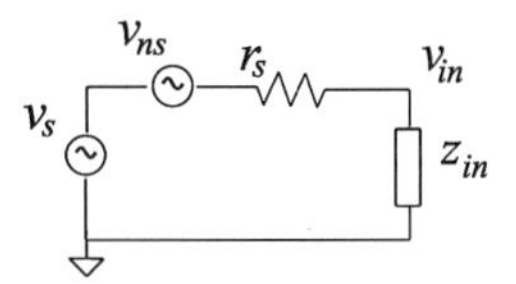

Fig. 4.19 Input loop with equivalent noise voltage signal

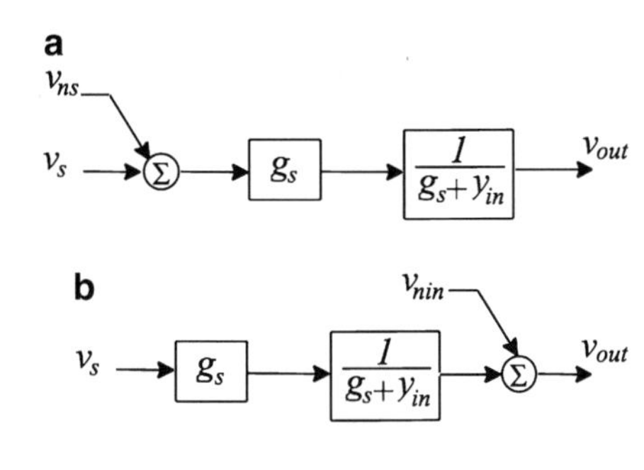

Fig. 4.20 (**a**) Flow graph for equivalent noise v_{ns} at source node and (**b**) flow graph with noise signal v_{ni} moved to output of the amp input loop as v_{nin}

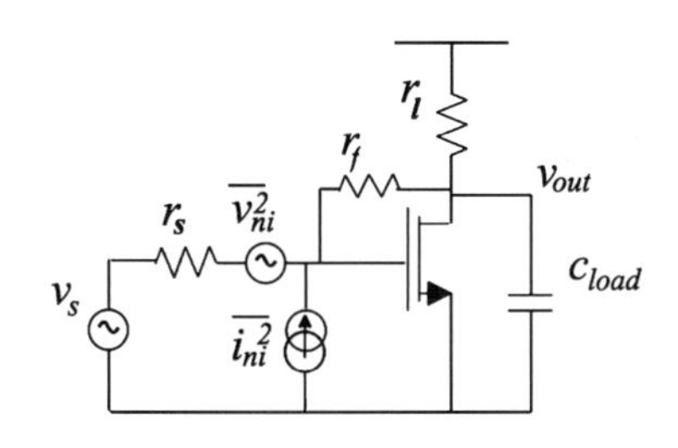

Fig. 4.21 Single transistor feedback amplifier with noise referred to input as two correlated noise sources

$$v_{\text{nin}} = \frac{g_{\text{s}}}{g_{\text{s}} + y_{\text{in}}} v_{\text{ns}} \tag{4.26}$$

This noise signal, v_{nin}, is now a function of the amplifier input admittance, y_{in}, and of the input conductance g_s. It is now moved back one block where it transforms to a shunt current as i_{nin} in Eq. (4.27).

$$i_{\text{nin}} = (g_{\text{s}} + y_{\text{in}}) \cdot v_{\text{nin}} \tag{4.27}$$

We now move the term proportional to g_{s} back one block in the flow graph converting it to a series noise voltage v_{ni} and leave the term proportional to y_{in} as a shunt current, i_{ni}, Eq. (4.28) to complete the transformation of output noise to input referred as the paired noise sources, Fig. 4.21. These sources are correlated through the factor v_{ns}.

$$\begin{aligned} i_{\text{ni}} &= y_{\text{in}} \cdot v_{\text{ns}} \\ v_{\text{ni}} &= g_{\text{s}} \cdot v_{\text{ns}} \end{aligned} \tag{4.28}$$

Noise Model for the CMOS Differential Stage

As we add complexity to the amplifier, we also add noise sources. A simple CMOS differential transconductance stage will have a noise schematic as shown in Fig. 4.21. Source-drain noise currents are added in shunt as before for all devices except for *N1* where we add two i_{nn1} sources, one into the drain and the other out of the source to produce the same circuit response and to simplify the analysis. This can be done to floating elements in general if it helps the analysis. Transistor *N2* is also a floating S-D device, and we do the same two-source equivalent to it in the analysis phase.

To find the total noise at the output node, we first find the net short-circuit current into the output node and then multiply by the output driving point impedance with both inputs at ac ground (Fig. 4.22). Assuming unity current mirror at the active pMOS load, we write for the output noise short-circuit current i_{nout} as Eq. (4.29).

$$i^2_{\text{nout}} = i^2_{\text{np2}} + i^2_{\text{np1}} + i^2_{\text{nn1}} + \left(\frac{i_{\text{nn1}}}{2} - \frac{i_{\text{nn1}}}{2}\right)^2 + i^2_{\text{n22}} + 0 \cdot i^2_{\text{nn3}} \tag{4.29}$$

$$v^2_{\text{nout}} = \left|\frac{1}{g_{\text{dp}} + g_{\text{dn}}}\right|^2 \cdot i^2_{\text{nout}} \tag{4.30}$$

The noise currents at the common source for *N1* and *N2* split and follow two paths to the output, one through *N2* with a gain of plus one-half and the other through *N1* and the pMOS current mirror, with minus one-half canceling out the *N2* path current. *N3* noise current also splits at the common source node into these two paths canceling each other as common-mode correlated excitation.

Normally we would divide by the DC gain of the cell to translate the output noise in Eq. (4.30) to the input. However, we know that the feedback net and load need to be taken into account to find the system behavior so that referring noise to the input at this point is premature. In the next section, we will add these elements to the analysis. The active single-in/single-out cell can be viewed as a two-port model having input and output noise signals in addition to the normal parameters (Fig. 4.23), where we use the *y*-parameters. Internal feedback is included with the

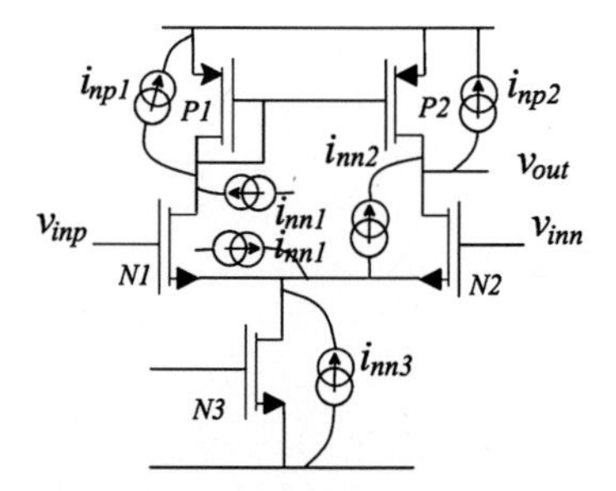

Fig. 4.22 CMOS differential transconductance stage with noise sources for each transistor

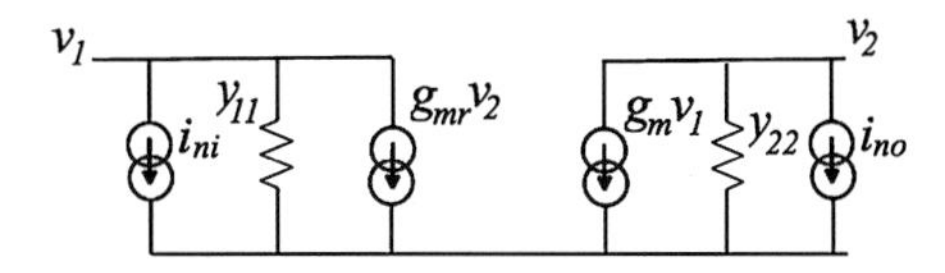

Fig. 4.23 Two-port model for general cells with noise added

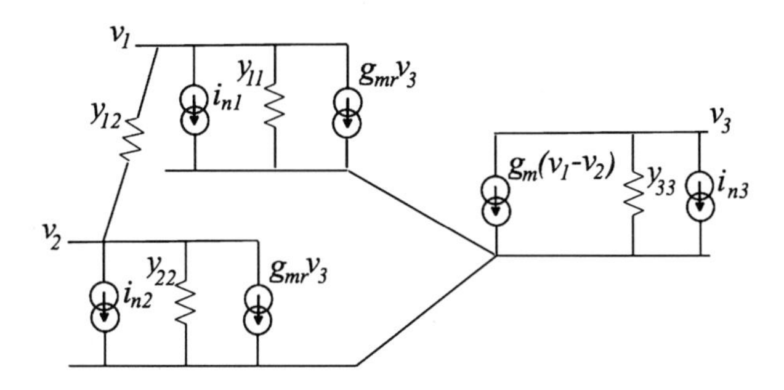

Fig. 4.24 Differential-in/single-out three-port small-signal model with noise

reverse transmission g_{mr} parameter. For the differential input stage, we need to add a port as in Fig. 4.24.

Adding a Load and Feedback Net

A feedback net can be modeled as a two-port model with added noise sources as shown in Fig. 4.23, making the feedback system the parallel combination of two of these models plus the source impedance input loop. While the models are needed to visualize the relations, we do not need to draw the composite system model as combinations of these. Instead we create the flow graph keeping the model form and relationships in our mind, directly from a schematic with added effective noise sources for each contributor (Fig. 4.25). For the amplifier, we have moved noise signals to the output as an output noise current i_{nampo}, and if we have a noise source directly at the input, we represent it there as i_{nampi}. The feedback net noise, resistor r_f, is shown as two i_{nrf} current sources, one into the output node and the other out of the amplifier inverting input node. Load noise is shown as i_{nload} and source and balancing impedance resistor noises as i_{nrs} and i_{nrsrf}.

Some of these noise currents are correlated. The feedback resistor r_f noise currents, $+-i_{nrf}$, are clearly correlated being the same physical source. Not clear at this level is the amplifier input noise current i_{nampi}. If the amplifier is modeled as that in Fig. 4.22, we may conclude that we do not have a noise component at the input. If the amplifier design included an internal resistive feedback path, or a noise model in series with the gate of the input devices, then we could have a nonzero input amplifier noise current i_{nampi}.

For this example, we simplify the differential gain stage model in Fig. 4.24 to be unidirectional ($g_{mr} = 0$) and remove the coupling between inputs, $y_{12} = 0$. Further,

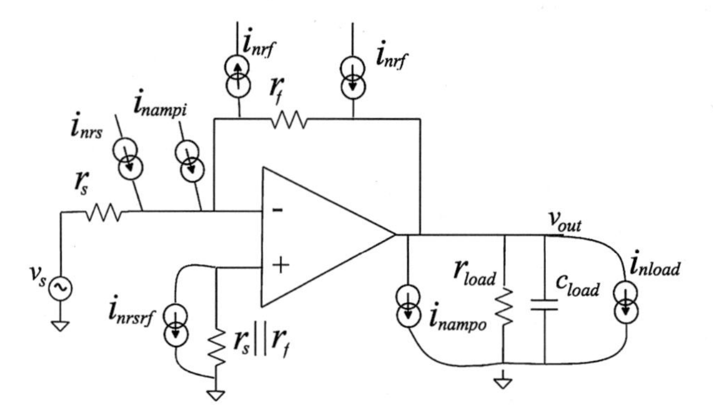

Fig. 4.25 Differential gain stage with load, source impedance, and noise contributors

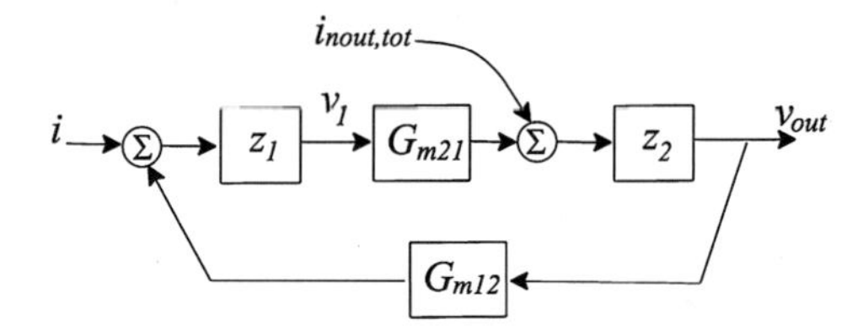

Fig. 4.26 Flow graph for amp, load, and feedback net of Fig. 4.26

we move the noise current at the non-inverting input directly to the inverting input node as equivalent due to the balanced local impedance being made equal by design. With these simplifications, we could proceed to directly create the system flow graph and perform additional noise moving manipulations on it.

Instead, let us look at the amplifier, feedback net, and load as a unit first. This will allow us to characterize the system before defining the source impedance. The flow graph looking into the inverting/feedback node is as shown in Fig. 4.26.

The graph parameters are given in Eq. (4.31) derived from the equivalent two-port models for the amplifier and feedback net plus the load impedance. $i_{nout,tot}$ represents the square root of sum of the squares of all the noise currents translated to the output node. This net noise current is moved to the output of the graph producing an output noise voltage, which we square and integrate in the frequency domain to obtain the variance of the output noise voltage, Eq. (4.32).

$$
\begin{aligned}
z_1 &= \frac{1}{g_\mathrm{f} + y_{11}} \\
z_2 &= \frac{1}{g_\mathrm{f} + y_{22} + g_\mathrm{load} + sc_\mathrm{load}} \\
G_\mathrm{m21} &= g_\mathrm{f} - G_\mathrm{mamp} \\
G_\mathrm{m12} &= g_\mathrm{f}
\end{aligned}
\tag{4.31}
$$

$$< v_{\mathrm{nout}}^2 >= \int_f \left| \frac{z_2}{1 - \mathrm{LP}} \right|^2 \cdot i_{\mathrm{nout}}^2 \mathrm{d}f \tag{4.32}$$

Solving for v_I in Fig. 4.26 collapses the flow graph loop as shown in Fig. 4.27a and, recognizing the first block as the input impedance for the composite cell of amplifier, load, and feedback net (Fig. 4.27b) shows the modified graph with z_{in} as the first block and the output noise voltage $<v_{nout}>$ summed at the output.

This transformed graph represents an equivalent view of the amplifier/feedback/load system that can be driven by the source and source impedance as shown in Fig. 4.28. We have characterized the feedback system independent of source drive so that we can attach it to a range of sources and better see the design consequences of the source on the overall system. Feedback net and load however are integral with the results.

We need to incorporate the noise from the input loop into our analysis and combine it with that of the amplifier, feedback net, and load. We do this by finding the output voltage due to the input noise source using the flow graph in Fig. 4.29, integrate the square of this response over frequency, and sum it with the square of $<v_{nout}>$ to find the total noise at the output $<v^2_{nout,tot}>$ (Eq. 4.33). It is necessary to do this at the system output as the frequency response of the system defines the filtering the system performs on the noise signal impacting the effective noise magnitude:

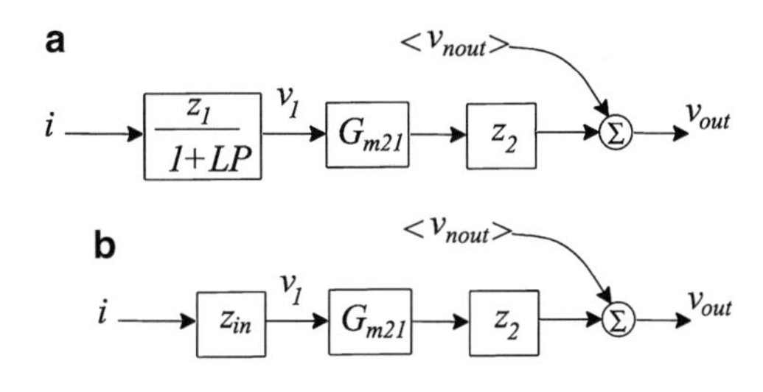

Fig. 4.27 Collapse of feedback flow graph loop for (**a**) v_I and (**b**) z_{in}

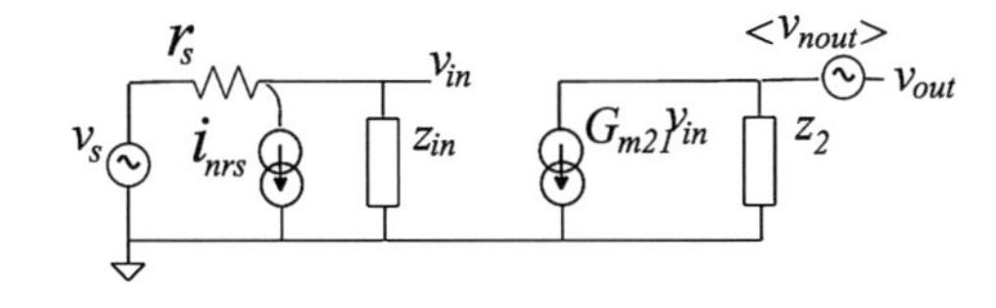

Fig. 4.28 Effective model for our amplifier/feedback/load and noise system after collapsing the loop to simplify the algebra

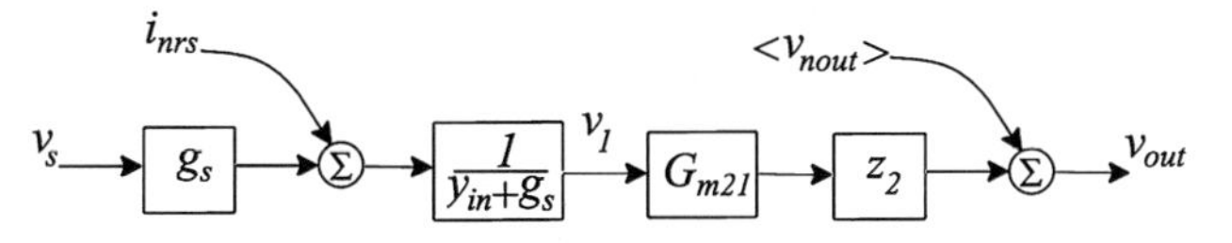

Fig. 4.29 System flow graph for the differential gain stage with feedback combining input source resistance and source noise signal

$$v^2_{\text{nout,tot}} = v^2_{\text{nout}} + v^2_{\text{ns}}$$
$$v^2_{\text{ns}} = \int_f i^2_{\text{nrs}} \left| \frac{1}{y_{\text{in}} + g_{\text{s}}} \right|^2 \cdot \left| G^2_{\text{m21}} \cdot z^2_2 \right| \mathrm{d}f \tag{4.33}$$

This net output noise voltage, $<v^2_{nout,tot}>$, can be moved back to the input v_n in Eq. (4.34) by dividing by the system gain that the signal will see, where it can be compared with the signal's root-mean-square voltage. For a low-pass gain stage, this gain is the dc gain of the system.

$$v_{\text{n}} = \frac{v_{\text{nout,tot}}}{G_{\text{m21}} \cdot z_2} \cdot (y_{\text{in}} + g_{\text{s}}) \cdot \frac{1}{g_{\text{s}}} \Big|_{\text{s}=0} \tag{4.34}$$

Equation Eq. (4.34) shows the output noise $v_{nout,tot}$ moved to the input where it is a voltage source in series with the source. It is a function of all of the system noise sources and the fully loaded amplifier system including the feedback net and is complete—we do not need the usual two-source representation.

To get the usual two-source equivalent noise representation, we start with the system flow graph with the total noise at the output, Fig. 4.30a, and move it toward the input by two blocks to get Fig. 4.30b. Interpreting this flow graph as Fig. 4.30c, we again see the floating noise source inside the small-signal model for the system. It is important to note that while we have divided the output noise by a quantity that looks like a gain, it is not the system gain as seen by the signal, and the noise voltage v_{n1} is not in the input loop at an equivalent position with the signal.

Moving v_{n1} in Fig. 4.30b back one more block as shown in Fig. 4.31a results in the shunt input current shown in Fig. 4.31b, labeled i_{n1} in Eq. (4.35). This noise signal is now outside of the system small signal model and, if we move it back across the g_{s} block, we get the v_n in Eq. (4.34), a single equivalent input noise

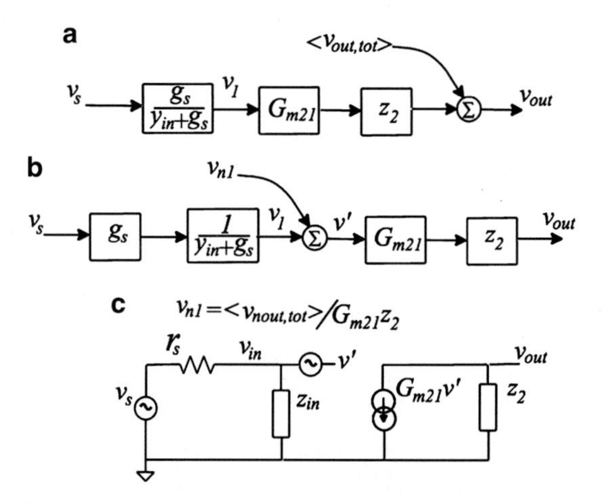

Fig. 4.30 (**a**) Flow graph with total noise at output, (**b**) flow graph with output noise moved to input node *1*, and (**c**) system equivalent circuit with floating noise voltage inside system equivalent circuit

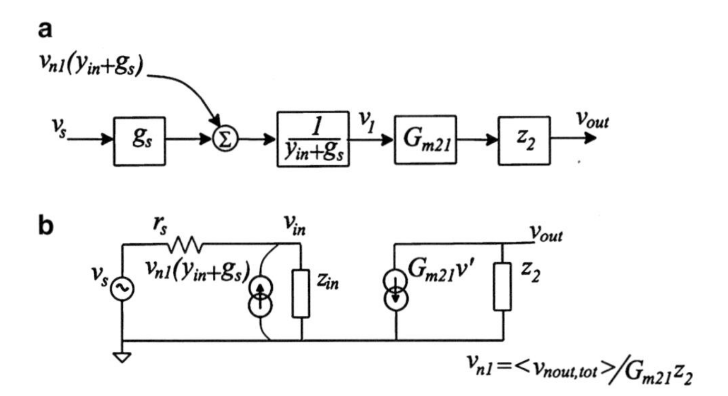

Fig. 4.30 (**a**) Flow graph with total output noise moved backward by two graph blocks where it appears as a shunt current at the input, and (**b**) equivalent small signal circuit for this transformation

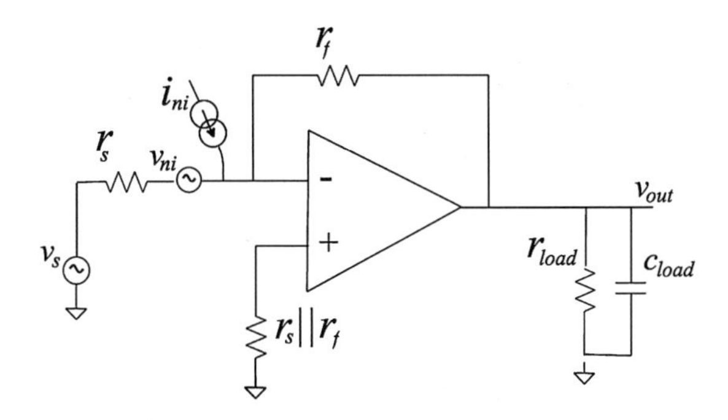

Fig. 4.32 Differential gain stage with all noise referred to input as two-source equivalent, correlated noise signals

voltage source having one term proportional to y_{in}/g_s and to $1/(G_m z_2)$, and the other term proportional only to the $1/(G_m z_2)$ factor. If instead we move only the term with the g_s factor of i_{n1} backward across the last g_s block in the system flow graph, we produce a series noise voltage while leaving a shunt noise current source at the amplifier input node, Fig. 4.32, where the noise sources are defined in Eq. (4.36) as the usual dual noise sources referred to the input.

$$i_{\mathrm{n1}} = \frac{v_{\mathrm{nout,tot}}}{G_{\mathrm{m21}} \cdot z_2} \cdot (y_{\mathrm{in}} + g_{\mathrm{s}}) \tag{4.35}$$

$$\begin{aligned} i_{\mathrm{ni}} &= \frac{v_{\mathrm{nout,tot}}}{G_{\mathrm{m21}} \cdot z_2} \cdot y_{\mathrm{in}} \\ v_{\mathrm{ni}} &= \frac{v_{\mathrm{nout,tot}}}{G_{\mathrm{m21}} \cdot z_2} \end{aligned} \tag{4.36}$$

The series voltage and shunt current noise signals are correlated through $v_{nout,tot}$.

Mismatch in Designs

A related design issue to noise is that of mismatch in devices. Mismatch is a statistical effect due to differences in local doping, patterning, and other manufacturing variables resulting in differences in device performance. Slight differences in effective threshold voltage between the differential input transistors results in an effective input offset voltage, for example. Here we focus on MOS devices that are wanted be made identical in a design—other elements follow the same pattern.

Consider the general MOS current relation dependency on two variables: threshold for the ith device, $v_{thi} = (V_{thi} - V_{th0})$, and beta, $\beta_i = (B_i - B_0)$, $B_0 = (\mu C_{ox}/2)W/L)$, where subscript '*0*' indicates the average for that parameter and subscript '*i*', the specific parameter for the ith device. While there will be some correlation between these variables, we will ignore them in this discussion as generally small. Taking the total differential of the drain current, I_D, with respect to these two variables, and dividing by I_D, gives the fractional change in drain current, Eq. (4.37), for square law operation.

$$\begin{aligned} I_{\mathrm{D}} &= I_{\mathrm{D}}(\beta, v_{\mathrm{th}}) = \beta \cdot F(v_{\mathrm{th}}) \\ \frac{\mathrm{d}I_{\mathrm{D}}}{I_{\mathrm{D}}} &= \frac{\partial I_{\mathrm{D}}(\beta, v_{\mathrm{th}})}{I_{\mathrm{D}}\partial\beta}\mathrm{d}\beta + \frac{\partial I_{\mathrm{D}}(\beta, v_{\mathrm{th}})}{I_{\mathrm{D}}\partial v_{\mathrm{th}}}\mathrm{d}v_{\mathrm{th}} \\ \frac{\mathrm{d}I_{\mathrm{D}}}{I_{\mathrm{D}}} &= \frac{\mathrm{d}\beta}{\beta} - \frac{g_{\mathrm{m}}}{I_{\mathrm{D}}}\mathrm{d}v_{\mathrm{th}} \\ \frac{\mathrm{d}I_{\mathrm{D}}}{I_{\mathrm{D}}} &= \frac{\mathrm{d}\beta}{\beta} - \frac{2\mathrm{d}v_{\mathrm{th}}}{(v_{\mathrm{gs}} - v_{\mathrm{th}})} \end{aligned} \tag{4.37}$$

With the variables uncorrelated, we find the fractional current variance as σ^2, Eq. (4.38), where *E*(.) is the expected value operation:

$$\begin{aligned} \sigma\left(\tfrac{\mathrm{d}I_{\mathrm{D}}}{I_{\mathrm{D}}}\right)^2 &= E\left(\tfrac{\mathrm{d}\beta}{\beta} - \tfrac{g_{\mathrm{m}}}{I_{\mathrm{D}}}\mathrm{d}v_{\mathrm{th}}\right)^2 \\ \sigma\left(\tfrac{\mathrm{d}I_{\mathrm{D}}}{I_{\mathrm{D}}}\right)^2 &= E\left(\tfrac{\mathrm{d}\beta}{\beta}\right)^2 + E\left(\tfrac{g_{\mathrm{m}}}{I_{\mathrm{D}}}\mathrm{d}v_{\mathrm{th}}\right)^2 \end{aligned} \tag{4.38}$$

The variance components of threshold voltage and for β have been found to have a term proportional to *1/WL* and a process-dependent constant one [3]. The process dependent Pelgrom variation term S_x is usually small and can generally be ignored.

$$\begin{aligned} \sigma\left(\tfrac{\mathrm{d}\beta}{\beta}\right)^2 &= \frac{A_\beta^2}{WL} + S_\beta^2 \\ \sigma(v_{\mathrm{th}})^2 &= \frac{A_{\mathrm{vth}}^2}{WL} + S_{\mathrm{vth}}^2 \end{aligned} \tag{4.39}$$

Mismatch Applied to a Simple Current Mirror

Current mirrors are found in most analog designs in some form or another. In its most basic form, a current is imposed onto a diode-connected transistor, drain tied to gate, with this node driving the gate of another transistor having the same source potential and of the same geometry (Fig. 4.33).

In this configuration the diode-connected transistor generates the particular v_{gs} required to produce the channel current imposed I_0. This v_{gs} is then used to drive a mirroring transistor having the same geometry. I_1 in Fig. 4.33 reflects the channel current response due to this gate drive. There will be a drain voltage effect that we ignore in this discussion, as it is a circuit effect rather than a device mismatch one and is handled using other circuit techniques. The mismatch is captured as i_{mm} in the mirroring branch.

Assuming $W = 2$ µm, $L = 2$ µm and using Pelgrom Coefficients $A_{vth} = 5.3$ mV-µm, $A_b = 1.04$ %-µm, representative values for 0.18 µm CMOS, for $I_0 = 2$ µA, and for $\beta_\square = 450$ µA/V^2, we calculate a mismatch current of about 16 % (Eq. 4.40).

$$\begin{aligned}\sigma\left(\tfrac{\mathrm{d}I}{I}\right)^2 &= \frac{A_\beta^2}{WL} + S_\beta^2 + \frac{4}{\left(v_{\mathrm{gs}} - v_{\mathrm{th}}\right)^2}\frac{A_{\mathrm{vth}}^2}{WL} + S_{\mathrm{vth}}^2 \\ &= \frac{0.0104^2}{2} + 0 + \frac{4}{0.047^2}\cdot\frac{0.0053^2}{2} + 0 \\ &= 0.00005 + 0.025 = 0.02505 \\ \sigma\left(\frac{\mathrm{d}I}{I}\right) &= 0.159\end{aligned} \tag{4.40}$$

This design will result in a large variation in the mirror current potentially causing yield loss. It can be improved by increasing the transistor lengths and widths keeping the ratio constant to maintain the same v_{gs}. To reduce this variation to 2 %, for example, will require $L = W = 35.7$ µm. Analog design costs area.

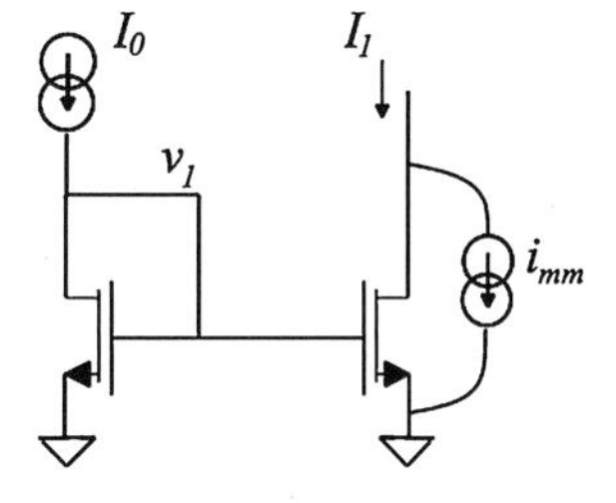

Fig. 4.33 Simple MOS current mirror showing mismatch current i_{mm}

A *1:M* Current Mirror

The *1:1* current mirror fits in well with the mismatch characterization generally available from the manufacturer. What about a mirror that is not *1:1*? For this we need to decouple the paired statistics to that of a single device. Given the variance for two devices, we know that it is composed of the individual device variances as the sum, so that the variance of a single device is half that of the mismatch generated variance, Eq. (4.41).

$$\begin{aligned} \sigma(v_{\text{th1}} - v_{\text{th1}})^2 &= E(v_{\text{th1}} - v_{\text{th2}})^2 \\ &= E\left(v_{\text{th1}}^2 - 2v_{\text{th1}} \cdot v_{\text{th2}} + v_{\text{th2}}^2\right)^2 \\ &= E\left(v_{\text{th1}}^2\right) + E\left(v_{\text{th2}}^2\right)^2 \\ \sigma(v_{\text{th}})^2 &= \frac{A_{\text{vth}}^2}{2 \cdot WL} \end{aligned} \tag{4.41}$$

Now, given the threshold voltage standard deviation for a single transistor, how do we inject it into our schematic? The threshold is an offset on the gate-substrate voltage signal and does not itself provide power to the circuit. It is in effect "inside" the model and appears floating as shown in Fig. 4.34a. We can tag the transistor with v_{th} to indicate that it has a statistical offset built in (Fig. 4.34b). The signal flow graph will look as in (c). Since the threshold variation is a dc signal, we do not need to include reactive elements such as the gate capacitance, as shown here. We set $s = 0$ in the analysis. For constant bias current I_O, $i_O = 0$ and v_I becomes as shown in Eq. (4.42).

$$\begin{aligned} v_1 &= v_{\text{th}} \cdot \frac{\frac{-g_{\text{m}}}{g_0}}{1 + \frac{g_{\text{m}}}{g_0}} \\ &\sim -v_{\text{th}} \end{aligned} \tag{4.42}$$

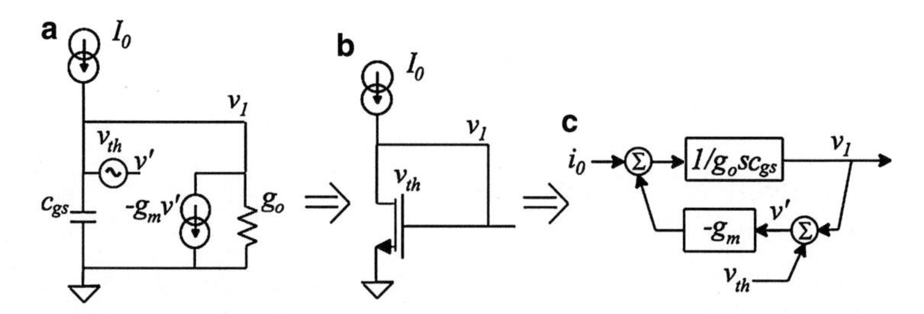

Fig. 4.34 Modeling the effect of v_{th} in the MOS transistor with gate tied to drain, the diode configuration

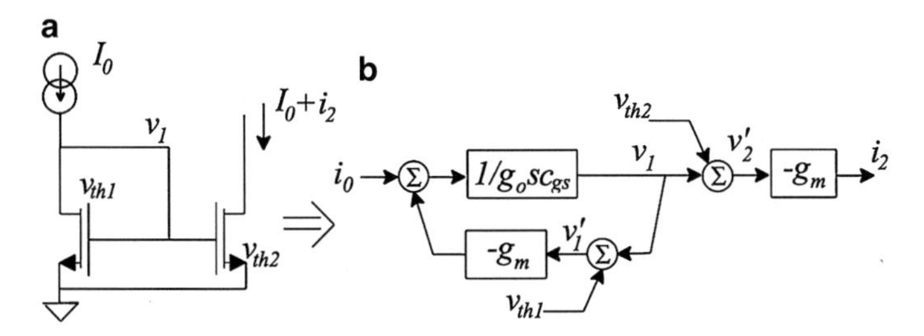

Fig. 4.35 (**a**) Adding a mirror transistor to the diode-connected transistor and (**b**) signal flow graph with internal v_{th} signals for 1:1 mirror

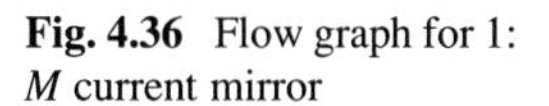
Fig. 4.36 Flow graph for 1: *M* current mirror

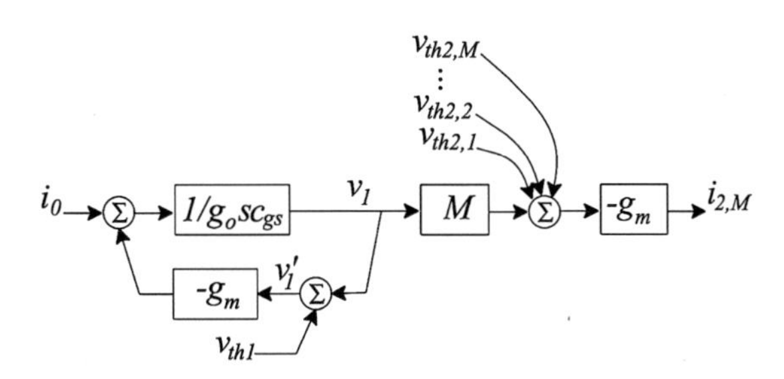

Adding a mirror transistor to the diode transistor in Fig. 4.34b having the same geometry results in Fig. 4.35, (a) the mirror schematic and (b) the signal flow graph for the full mirror using the v_{th} floating source internal to the small-signal transistor model.

Graph and statistic algebra produces the standard deviation for i_2 (Eq. 4.43) repeating the result from Eq. (4.41).

$$\begin{aligned} < i_2^2 > &= E[(v_{\text{th1}} + v_{\text{th2}})g_{\text{m}}]^2 \\ &= \left[E\left(v_{\text{th1}}^2\right) + E\left(v_{\text{th2}}^2\right)\right]g_{\text{m}}^2 \\ &= 2\left[\sigma^2\left(v_{\text{th}}\right)\right]g_{\text{m}}^2 \end{aligned} \tag{4.43}$$

Returning to our *1:M* mirror problem, we see that we have set up the mechanics for this and for general circuit mismatch problems. Making the mirror transistor *M* parallel unit devices in Fig. 4.35a, we get the *1:M* mirror. The diode device adds a variation onto the gate node used by *M* mirror transistors in parallel adding to the output current i_2. In addition, each of the *M* output mirror devices contributes its current variation due to its particular threshold voltage. We add *M* threshold deviations to *M* times the diode threshold variation as shown in the flow graph in Fig. 4.36 and statistically combine the results (Eq. 4.44).

$$i_{2,M} = \left(M \cdot v_{\text{th1}} + \sum_{j=1}^{M} v_{\text{th2},j}\right) g_{\text{m}}$$

$$\begin{aligned}\sigma^2(i_{2,M}) &= E\left[\left(M \cdot v_{\text{th1}} + \sum_{j=1}^{M} v_{\text{th2},j}\right) g_{\text{m}}\right]^2 \\ &= \left[M^2 E\left(v_{\text{th1}}^2\right) + ME\left(v_{\text{th}}^2\right)\right] g_{\text{m}}^2 \\ &= \sigma^2(v_{\text{th}}) \cdot \left(M^2 + M\right) \cdot g_m^2\end{aligned} \tag{4.44}$$

Equation (4.44) gives us the variance of the output current. To get the fractional variance and the fractional standard deviation for this *1:M* current mirror, we divide by the square of the ideal output current $(M \times I_0)^2$, Eq. (4.45).

$$\begin{aligned}\sigma^2\left(\frac{i_{2,M}}{MI_0}\right) &= \frac{\sigma^2(v_{\text{th}}) \cdot \left(M^2 + M\right) \cdot g_{\text{m}}^2}{M^2 I_0^2} \\ &= \frac{4\sigma^2(v_{\text{th}})\left(M^2 + M\right)}{M^2\left(V_{\text{gs}} - V_{\text{th}}\right)^2} \\ &= \frac{2A_{\text{vth}}^2\left(1 + \frac{1}{M}\right)}{WL\left(V_{\text{gs}} - V_{\text{th}}\right)^2}\end{aligned} \tag{4.45}$$

Here we use Eq. (4.41) replacing the standard deviation for a single transistor threshold voltage with that of the Pelgrom coefficient (threshold only) for a pair of transistors. The *1:M* current mirror expected range expressed as a fraction or as a percent (multiply by 100) is improved by (1) increasing transistor area *WL*, (2) increasing the gate overdrive $(v_{gs} - v_{th})$, and (3) increasing *M*. This latter effect is due to the averaging effect of the multiple parallel devices.

Mismatch Offset in a Differential Gain Stage

To find the input offset voltage due to device pair mismatch, consider the differential gain stage in Fig. 4.37. Current sources i_{mp12} and i_{mn12} are added as the p- and n-pair mismatch currents. The stage is biased with the negative input v_{inn} at a DC voltage and the positive input v_{inp} at this DC voltage plus a small-signal v_s. The circuit signal flow graph is shown in Fig. 4.38.

Moving the mismatch currents backward to the input as was done with the noise signals, we divide each by the g_{mn}, and expressing the result as a function of the threshold sigma, we get Eq. (4.46). The input transistor pair threshold mismatch

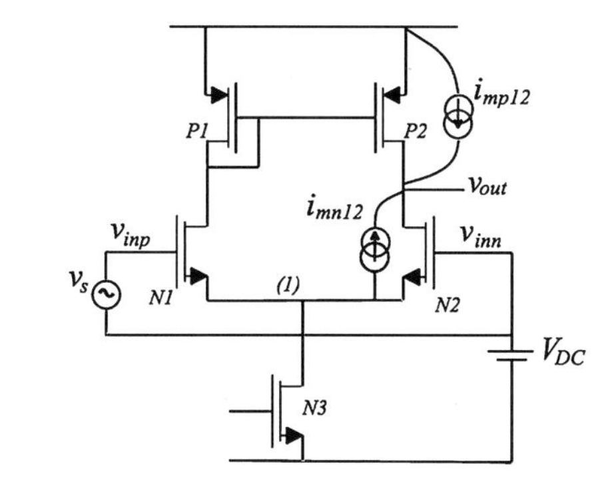

Fig. 4.37 Differential gain stage with mismatch currents included for the P-load and the N-differential input pairs

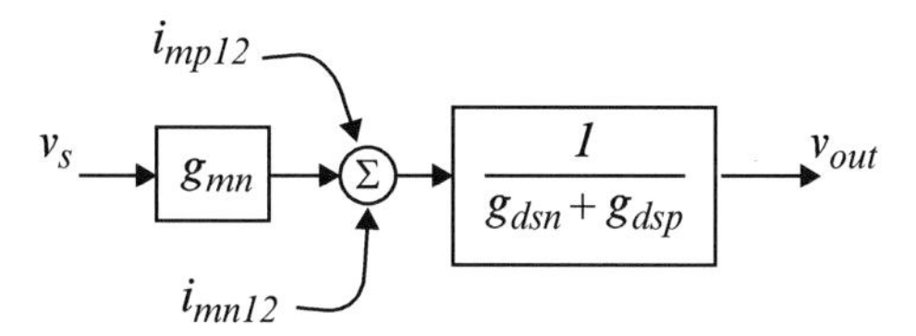

Fig. 4.38 Signal flow graph for differential stage with mismatch transistor pair currents

voltage variance is moved directly to the input while that of the load transistors is scaled by square of the ratio (g_{mp}/g_{mn}), the ratio of the load device transconductance to that of the input differential pair. *Note that this input offset voltage is again inside the small-signal model for the amplifier* as we found for the current mirror in Fig. 4.34 although in this case the source impedance is assumed zero so that it does not make a difference, and the source and the offset source do not draw current.

$$\begin{aligned} < v_{\mathrm{in,off}}^2 \geq &< \frac{i_{\mathrm{mn12}}^2}{g_{\mathrm{mn}}^2} + \frac{i_{\mathrm{mp12}}^2}{g_{\mathrm{mn}}^2} > \\ &= \sigma_{\mathrm{vthn12}}^2 + \sigma_{\mathrm{vthp12}}^2 \frac{g_{\mathrm{mp}}^2}{g_{\mathrm{mn}}^2} \end{aligned} \tag{4.46}$$

Offset Voltage in a Bandgap Reference Voltage Circuit

We now look at the effect of an input offset voltage in an amplifier in a bandgap reference voltage generator cell. A common configuration is shown in Fig. 4.39.

This cell balances one signal that decreases with increasing temperature with one that increases. Briefly, feedback maintains the amplifier inputs at the voltage across the unit bipolar transistor with base collector shorted to behave as a diode, device *1* on the left branch. Device *M* consists of *M* copies of device *1* wired in parallel. With the diode currents equal, the difference between their forward voltages is imposed across resistor r_1. This voltage, given as the well-known

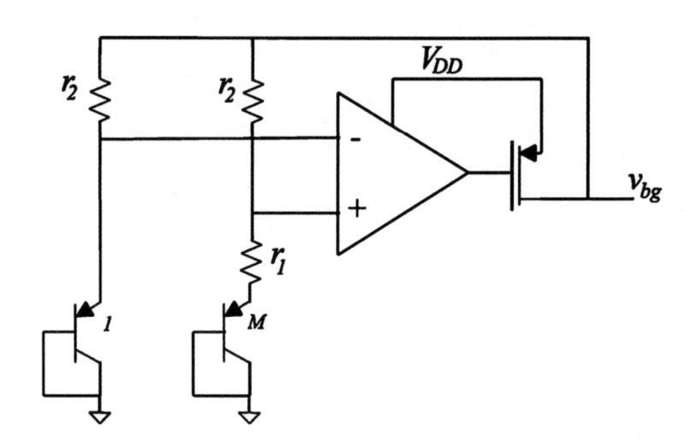

Fig. 4.39 Bandgap voltage reference cell

relation in Eq. (4.47), is proportional to absolute temperature, PTAT, while the diode voltage decreases with temperature and is called CTAT or "complementary to absolute temperature".

$$\Delta v_{\mathrm{be}} = \frac{kT}{q}\ln(M) \tag{4.47}$$

The output of the amplifier is applied to the top node common to the two r_2 resistors making the voltage across them equal. This makes the currents in the two branches equal. The top node voltage, v_{bg}, is given in Eq. (4.48).

$$v_{\mathrm{bg}} = v_{\mathrm{be1}} + r_2\frac{kT}{q}\ln(M)\frac{1}{r_1} \tag{4.48}$$

By proper sizing of the two resistors, v_{bg} can be made to be relatively stable with temperature at about 1.2 V.

A single-stage differential input amplifier such as that in Fig. 4.37 will have an input offset voltage as given in Eq. (4.46). How will this offset affect v_{bg}?

To answer this we need to complete the design. We start by selecting *M*. This value is generally between 8 and about 35. We use numbers such that the diodes can be placed into a rectangle with the unit device near the center to minimize spatial variations. For $M = 8$, the number of diodes is *9* so that a 3×3 diode array is used. Here we will use *15* making the diode array 4×4. This design parameter defines the voltage across r_1 in our voltage reference design.

Next we select the branch currents. This would be based on other inputs such as transient behavior, power limits, and cell area. Let us arbitrarily select *1* μA at *27* °C per diode branch.

Given these two variables, we can define r_1 as in Eq. (4.49).

$$\begin{aligned} r_1 &= \frac{kT}{q}\ln(M)\frac{1}{I_{\mathrm{d}}} \\ &= \frac{0.026 \cdot \ln(15)}{1e - 6} = 70.4\ \mathrm{k}\Omega \end{aligned} \tag{4.49}$$

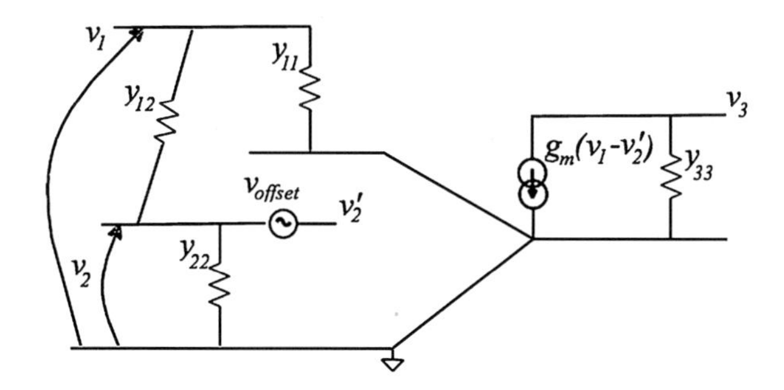

Fig. 4.40 Small-signal equivalent model for differential-in/single-out amplifier with offset voltage referred to input inside small-signal model

With the current and temperature specified, we can obtain the diode forward voltage $v_{be}(27C)$ and use this plus the nominal bandgap voltage of 1.2 V to obtain r_2 (Eq. 4.50).

$$\begin{aligned} r_2 &= \frac{1.2V - v_{\mathrm{be}}(27C)}{I_{\mathrm{d}}} \\ &= \frac{1.2 - 0.6}{1e - 6} = 600\,\mathrm{k}\Omega \end{aligned} \tag{4.50}$$

With the bandgap design roughed in, we next insert the offset voltage as a voltage source v_{offset} inside the small-signal model at one of the inputs, here at the positive input as shown in the flow graph in Fig. 4.40. With this model in mind, from the bandgap circuit in Fig. 4.39, we write directly for v_{bg}, Eq. (4.51), where we use the DPI law, $v_{out} = i_{sc} \times dpi_{out}$.

$$\begin{aligned} v_{\mathrm{bg}} &= v_{\mathrm{offset}} \cdot G_{\mathrm{mo,voff}} \cdot \mathrm{DPI}_{\mathrm{out}} \\ \mathrm{DPI}_{\mathrm{out}} &= \frac{R_1||R_M}{1 + (\alpha_M - \alpha_1) \cdot G_{\mathrm{mo,voff}} \cdot R_1||R_M} \\ R_1 &= r_2 + r_{\mathrm{d}} \\ R_M &= r_2 + r_1 + r_{\mathrm{d}} \\ \alpha_M &= \frac{r_1 + r_{\mathrm{d}}}{r_2 + r_1 + r_{\mathrm{d}}} \\ \alpha_1 &= \frac{r_{\mathrm{d}}}{r_2 + r_{\mathrm{d}}} \\ v_{\mathrm{bg}} &\sim v_{\mathrm{offset}} \frac{1}{\alpha_M - \alpha_1} \end{aligned} \tag{4.51}$$

Here, $G_{mo,voff}$ is the "transconductance" at the output due to excitation $v_2{}'$, R_1 is the branch resistance of the first branch (left) with the single diode, from the v_{bg} node to ground, and R_M is the branch resistance for the branch with M diodes. The α's are voltage division factors from node v_{bg} to the amplifier inputs for each branch. The diode small-signal resistance, r_d, is the same for both diodes being a function of forward current and temperature. Because we break the analysis down into smaller parts, we can write these results down by inspection. The term added to 1 in the denominator is much larger than 1 allowing us to drop the 1 at the frequency of interest (DC). We then cancel common factors top and bottom leaving the gain at the output node v_{bg} due to v_{offset} equal to the reciprocal of the difference in the branch voltage division factors. For $M = 15$, the gain due to v_{offset} is 10.3. Decreasing M to *8*, we get *13.1*, while increasing to *24* results in an offset gain of *8.9*.

Summary

This chapter introduced graph manipulations allowing us to move signals from where they appear in a circuit to another position as an effective signal. We move noise signals to cell inputs where we can compare these to the system signal for signal-to-noise calculations and mismatch signals to inputs to compare with signal levels to determine if we can meet a quantization specification, for example. Noise and mismatch signals are statistical signals. We use statistical algebra to quantify and combine with other similar signals in the circuit.

In the process of moving signals to a cell input, we found that at the cell transition, we create a floating voltage source inside the cell's small-signal model that is not in a loop. It draws no current. This is a common misconception in the literature and leads to the correction of creating a mathematical balancing current source. With the flow graph manipulations, we see this floating source and treat it accordingly to maintain noise equivalence at all times and we do not need to introduce a correction. In mismatch analysis, this same floating source appears. If we erroneously place these voltage sources outside of the cells at this transition, we cause them to supply current in the input loop that should not be there, producing errors dependent on loop impedance levels. In the noise analysis section, we saw that a given element, a resistor, for example, may introduce two noise signals into the circuit, one at each of its end points. These noise signals are fully negatively correlated, one entering the circuit and the other exiting. Knowing this, we maintain correlation throughout the process of translation and can properly combine these correlated signals at their final points. Similarly, we saw that once we combined all noise signals at a cell's output, we can move this net noise to the input where we could choose to split its representation into the two-source equivalent. Following this process we see how to maintain correlation between these two sources representing the input-referred noise.

We applied the statistical definition and translations of noise and mismatch signals to a few sample circuits to develop the methodology. Noise in a low-pass filter was shown to produce a standard deviation dependent only on the capacitor and temperature, plus Boltzmann constant. Noise in simple amplifiers was analyzed. In mismatch we examined two current mirror structures, a *1:1* mirror and a *1:M* current mirror. We saw how to apply the mismatch analysis to amplifiers to find the effect of amplifier input offset voltage in a bandgap reference voltage cell to its output.

References

1. A. Ochoa, *Translating Noise Signals in Linear Circuits. 42nd Midwest Symposium on Circuits and Systems*, vol. 2 (IEEE, Las Cruces, NM, 1999), pp. 955–958
2. B. Razavi, *Design of Analog CMOS Integrated Circuits* (McGraw Hill Publishing, Singapore, 2000)
3. M.J.M. Pelgrom, A.C.J. Duinmaiger, A.P.G. Welbers, Matching properties of MOS transistors. IEEE J. Solid-St. Circ. **24**, 1344–1349 (1989)

Chapter 5
Feedback: A Unified Treatment of Feedback in Analog Circuits

Abstract Feedback in analog circuits is an important design tool providing a means to produce better control of circuit behavior by combining a part of the system output with the system input and processing this combination. This process trades off amplifier gain for improved behavior stabilizing a design to variations in amplifier parameters and to temperature. Feedback can also lead to instability in a design causing ringing and unwanted oscillations. The analysis of analog circuits containing feedback has long been confusing due to incomplete definitions of fundamental terms such as loop gain, open-loop gain, and "port impedance with feedback zeroed," z_0', and is complicated by loading and by bi-directionality signal flow in elements. In this chapter, following the driving point impedance-signal flow graph approach, a unified analysis process is developed, the Z-Method, that can be translated into the Black's block diagram to properly identify the feedback analysis parameters and definitions. This approach fully accounts for bi-directional properties in the amplifier and in the feedback net as well as loading effects. Feedback is de-mystified.

Introduction

Devices used in the design of analog circuits are not ideal. They suffer from variations in property values that also change with temperature. An integrated circuit resistor, for example, composed of a diffused region in the substrate may vary in value by as much as $+-40\ \%$ and change with temperature. Active devices also will have temperature-dependent model parameters specified, making open-loop designs highly variable in input-output transfer functions. Designing analog circuits with components having such variation would not be practical except for feedback. Feedback is a design tool where a portion of a gain cell's output is combined with its input, leading to a significant improvement in the behavior of a design. H. Black [1] first described this process using a block diagram as depicted in Fig. 5.1. An inverting gain stage has a portion of its output returned to the system input where it is added to the input signal and the combined signal is reprocessed by the system. Using basic flow graph algebra we write for v_{out} as a function of v_{in},

A. Ochoa, *Feedback in Analog Circuits*, DOI 10.1007/978-3-319-26252-9_5

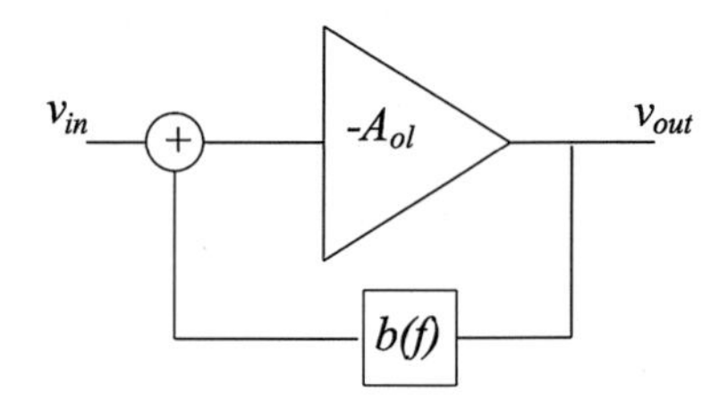

Fig. 5.1 Block diagram showing the basic blocks of feedback

Eq. (5.1) where it is assumed that the loop product is large compared to one. In this model where blocks are ideal and unilateral, the loop product is the same as the loop gain.

$$v_{\text{out}} = \frac{-A_{\text{ol}}(f)}{1 + b(f) \cdot A_{\text{ol}}} v_{\text{in}} \sim \frac{-v_{\text{in}}}{b} \tag{5.1}$$

The feedback factor "$b(f)$," in Eq. (5.1) can be designed to be a ratio of like passive devices and as such, variations in value as well as changes with temperature cancel leaving a stable system gain. The negative gain in the amplifier, $-A_{ol}(f)$, is needed so that the returned signal subtracts from the input signal, and feedback is *negative*. Note that the return signal here is *added* to the input signal, in contrast to its general definition in the feedback literature where it is subtracted, so that all phase shift information is contained in the design blocks themselves where it exists. This use eliminates the *180°* reference phase imposed by the use of subtraction found in the general feedback literature going back to Black [1].

We see from Eq. (5.1) that for a large loop product (loop gain), the transfer function is not sensitive to variations in the amplifier gain, A_{ol}. To quantify this we can take the fractional change in the output response due to the fractional change in the amplifier gain, A_{ol}, holding the feedback factor b constant.

$$\frac{\delta(v_{\text{out}})}{v_{\text{out}}} = \frac{1}{1 - b \cdot A_{\text{ol}}} \cdot \frac{\delta(A_{\text{ol}})}{A_{\text{ol}}} \tag{5.2}$$

Equation (5.2) shows that the impact of the fractional variation in the amplifier gain on the output voltage response is reduced by the loop gain. An open-loop design having a nominal gain of *40 dB* with a +–*6 dB* gain variation can be improved by using feedback with an amplifier gain of *80 dB* to have a variation of *–34 dB*. The added amplifier complexity of increasing the gain from *100* to *1000* improves the system gain variation from *50 %* to about *1.5 %*.

This introduction oversimplifies feedback. Ideal blocks are unidirectional and do not have loading effects. Real systems are not ideal—loading and bilateral effects impact the analysis and the performance of a design. Phase in real systems is not the constant *180°* attributed to the amplifier but varies with frequency. In this simplified model, the product $b(f) \times A_{ol}(f)$, what we will call the *loop gain*, will vary in both phase and in magnitude with frequency. Should this loop gain have near *180°* phase change for a net *360°* phase change at the frequency where its magnitude drops to *one*, the system will become unstable exhibiting ringing and may even sustain oscillations. The design of feedback systems deals with all of these issues including variation due to manufacturing and with the effects of temperature changes on the behavior of devices. Loop gain and properties of loop gain will be introduced as monitors and characterization of feedback designs, developed to include loading and bi-directional signal transfer across design blocks.

Loop Gain in Feedback Design

We have already analyzed a number of circuits with feedback in the preceding chapters and did not comment much on this feature. We solved for transfer functions for single-transistor feedback design, for the inverting gain stage using a general-purpose amplifier, we analyzed a bandgap voltage reference design for mismatch and other introductory circuits, not caring that the design had feedback. Indeed a property of the DPI/SFG methodology is that it uses sub-circuits in the analysis process, circuits which can be selected to have feedback disabled. Why do we need a chapter on feedback at this point? While we have been able to analyze feedback circuits directly and easily, we have not set up the analysis with feedback design in mind. The transfer functions obtained have "looked" like Black's relation in Eq. (5.1), but we have been careful not to call the term in the denominator subtracted from one, the "loop gain." Instead we called it the "loop product." Algebra will produce results in this form, but unless we design the analysis to produce the loop gain, we will not know that that is what we obtain. And it is the loop gain that carries loop stability information, phase, and magnitude that can be used to specify design stability.

In order to create the loop gain function, we need to identify a signal that circulates through the loop. Simply doing algebra on a small-signal system model is not likely to be appropriate [2, 3]. Bode [4] introduced the "return ratio" method using the response of a controlled source in an active device as the signal to be tracked. Replacing the controlled source in an active device with a unit source, the ac response at the controlled source that remains coupled to its controlling node, as shown in Fig. 5.2, is the return ratio *RR*.

This approach effectively breaks the loop at the controlled source and drives the loop from this point. The return ratio method is well defined and intuitively produces the loop gain. In a multiple transistor design, any of the active devices in the loop can be used, as each will produce the same return ratio. The problem with its use arises when we try to simulate a design since accessing a controlled source in modern models is difficult. Hand analysis is possible using simple models, but bench measurements are simply not possible [5]. The return ratio is the basis for the

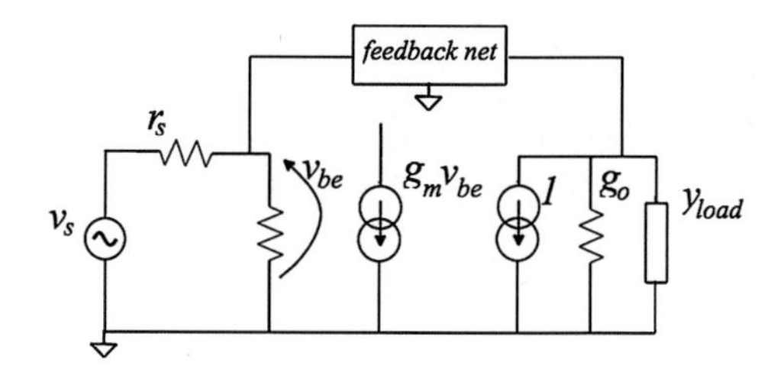

Fig. 5.2 Return ratio setup for a single-transistor feedback amplifier

popular “open-loop” method for hand analysis and simulations, where a high impedance point in the loop is used as the break location for a voltage loop gain drive signal, a low impedance point for a current drive [2]. Care must be taken to properly terminate loop breaks as well as to maintain DC circuit conditions on both sides of the break, and neither generally accounts for reverse transmission in the loop.

The Z-Method for Feedback Analysis

The z-method [6] looks at a node in the outer feedback loop and evaluates the port impedance at that point to develop the system loop gain function as one of the port impedance components. It produces the system loop gain function that is free from all of the problems enumerated above for the various methods generally used such as the popular “open-loop” method and the return ratio and includes a reverse transmission loop gain term that is likely to be ignored. In addition, with this approach, we properly develop the A_{ol} factor as being dependent on its full environment, the load, the feedback net, and even the input source impedance.

The basic relation of the DPI method is that the voltage response at a node is the product of the short-circuit current at that node with the driving point impedance looking into that node. The DPI/SFG method has been described as multiple uses of this statement such that two or more sub-circuits are analyzed with the results interpreted as a flow graph. Using this approach feedback circuits are easily analyzed, but loop gain is not specifically a part of the result. To find the loop gain for a feedback circuit, we start by selecting any node in the outer loop. Consider the feedback system shown in Fig. 5.3. Arbitrarily taking the output node as the node for analysis, we solve for the DPI variables, the short circuit current i_{sc} and the port driving point impedance dpi_{out} for this node.

$$i_{\mathrm{sc}} = G_{\mathrm{mos}} \cdot v_{\mathrm{s}} \tag{5.3}$$

$$G_{\mathrm{mos}} = g_{\mathrm{s}} \cdot \frac{1}{g_{\mathrm{s}} + y_{11\mathrm{f}} + y_{11\mathrm{a}}} \cdot (y_{21\mathrm{f}} - G_{\mathrm{ma}}) \tag{5.4}$$

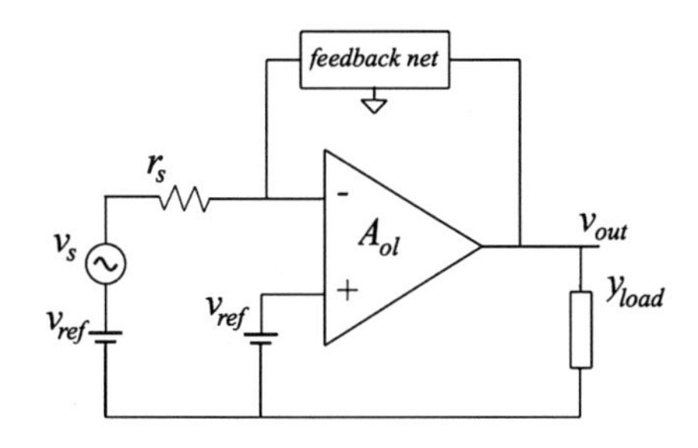

Fig. 5.3 Feedback system used for demonstration of the z-method for feedback analysis

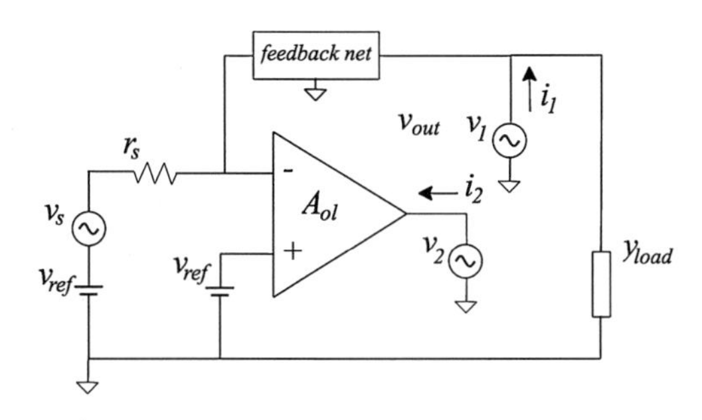

Fig. 5.4 Feedback system with output node split for finding output DPI

To find the output *DPI*, we split the node at the amplifier output into two nodes, apply a voltage source at each of the split nodes (Fig. 5.4) and find the total current introduced into the combined nodes. The load branch in this figure is tied to the split node at the feedback net arbitrarily. It could have been tied to the amplifier output as well. We name the sources v_1 and v_2 to tag the currents with the driving source, making them equal to v_{out} at the end of the analysis. Each of the source currents will have two components, a self-current, i_{xx}, due to its own source, and a cross component, i_{xy}, due to the other source, Eq. (5.5).

$$\begin{aligned} i_{\text{total}} &= i_1 + i_2 \\ &= i_{11} + i_{12} + i_{21} + i_{22} \end{aligned} \tag{5.5}$$

Separating each current into two parts, one arising from the source attached directly to a branch and the other arising from the source attached to the other branch, allows us to separate that portion of the total current that flows to ground, i_{xx} or "self-current," from that portion that traverses the loop, i_{xy}, or "crosscurrent." With this identification we can process the signals and maintain this distinction. Sources v_1 and v_2 are now made equal to v_{out}. Multiplying and dividing each current component by v_{out} produces an admittance factor for the self-currents and a transconductance factor for the crosscurrents, Eq. (5.6), interpreted as the small-signal equivalent circuit in Fig. 5.5 and the flow graph in Fig. 5.6.

$$i_{\text{total}} = (y_{11} + y_{22} + g_{\text{m21}} + g_{\text{m12}}) \cdot v_{\text{out}} \tag{5.6}$$

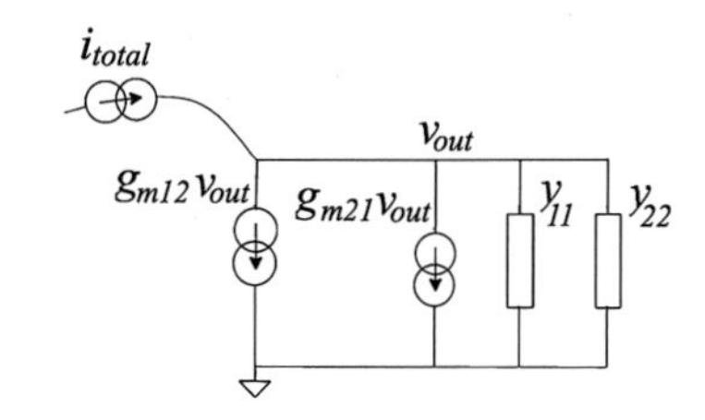

Fig. 5.5 Small-signal equivalent circuit for dpi_{out} for the feedback system shown in Fig. 5.4

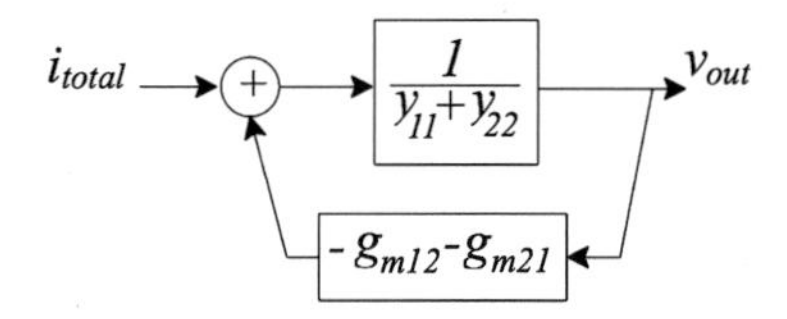

Fig. 5.6 Output port driving point impedance flow graph for algebra in Eq. (5.7) showing different roles between self-currents and crosscurrents

The transfer function v_{out}/i_{total} is the desired dpi_{out}, Eq. (5.7):

$$\begin{aligned} \mathrm{dpi_{out}} &= \frac{\frac{1}{y_{11}+y_{22}}}{1-\frac{1}{y_{11}+y_{22}}\cdot(-g_{m21}-g_{m12})} \\ &= \frac{z_0'}{1-\mathrm{LG}} \\ z_0' &= \frac{1}{y_{11}+y_{22}} \\ \mathrm{LG} &= \frac{1}{y_{11}+y_{22}}\cdot(-g_{m21}-g_{m12}) \end{aligned} \tag{5.7}$$

Equation set (5.7) shows z_0' as the port impedance with the loop gain zeroed and the system loop gain as the loop product of the flow graph in Fig. 5.6 where we have differentiated between currents that flow to ground from those that flow in the loop. To "zero" the loop gain we disable the transconductors, the crosscurrents in the system, by opening the graph loop. This action is performed in the math domain and not in the actual circuit where simply opening the loop upsets nodal impedances. It is important to note that this analysis accounts for signals flowing in both directions—one in the direction of the amplifier input to output and the other in the reverse direction. This symmetry is necessary to include reverse transmission through both the amplifier and the feedback net. The modeling is complete accounting for loading and for reverse transmission through elements.

Let us take another look at the formulation of LG in Eq. (5.7). Multiplying LG by an arbitrary v/v transforms LG from a ratio of transconductance to admittance to a ratio of currents—*minus the sum of the cross-currents to the sum of the self-currents* [6], Eq. (5.8). This loop gain definition accounts for loading and reverse transmission through the loop. Further, it is not restricted to the use of a controlled source of an active element in the design but is a general result of two-port modeling of the amplifier and feedback net plus load and is in a form lending itself to hand analysis as well as to computer simulations and to bench measurements.

$$\mathrm{LG} = \frac{-(i_{21} + i_{12})}{i_{11} + i_{22}}. \tag{5.8}$$

General Cell Transfer Function

Let us continue with the development of the transfer function analysis for the gain stage with feedback in Fig. 5.3. Coupling the results for the short-circuit current, Eq. (5.4), with the output node *DPI*, Eq. (5.7), we have a complete transfer function analysis following the *DPI/SFG*, single-output node impedance approach, Fig. 5.7.

Recall that we have combined parameters—load, amplifier, and feedback net, into the graph parameters *as seen at the node of analysis*. G_{mos}, as we have specified the node for analysis to be the output node, does not involve the system load as seen in Eq. (5.4) and repeated here as Eq. (5.9).

$$G_{\mathrm{mos}} = g_{\mathrm{s}} \cdot \frac{1}{g_{\mathrm{s}} + y_{11\mathrm{f}} + y_{11a}} \cdot (y_{21\mathrm{f}} - G_{\mathrm{ma}}) \tag{5.9}$$

The first product forms a voltage divider between the signal source v_s and the inverting input node to the amplifier (under the output shorted condition). The terms in parenthesis represent the forward transconductance properties for the feedback net (y_{21f}) and for the amplifier ($-G_{ma}$). Note that it is completely general, that the amplifier effective transconductance may include an internal feedback path or internal input-output shunt conductance should the amplifier not be fully unidirectional. By replacing G_{mos} with Eq. (5.9) in the flow graph in Fig. 5.6, we get the graph in Fig. 5.8, where z_0' and LG are given in Eq. (5.7).

Next, move the amplifier transconductance into the loop, Fig. 5.9, where it goes into both loop blocks, in the numerator for the forward block and in the denominator for the reverse block, keeping the loop gain the same as before this graph manipulation.

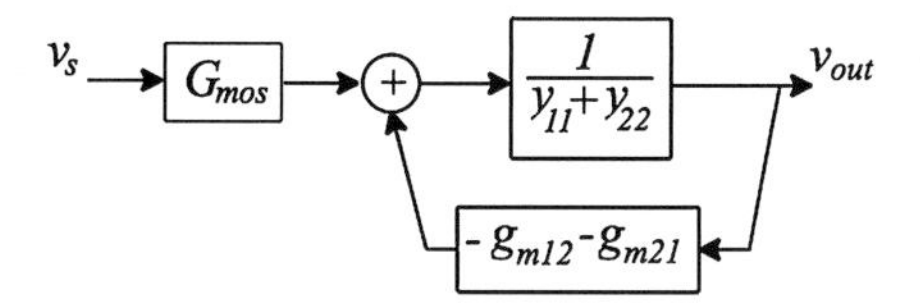

Fig. 5.7 Flow graph for feedback amplifier cell shown in Fig. 5.4

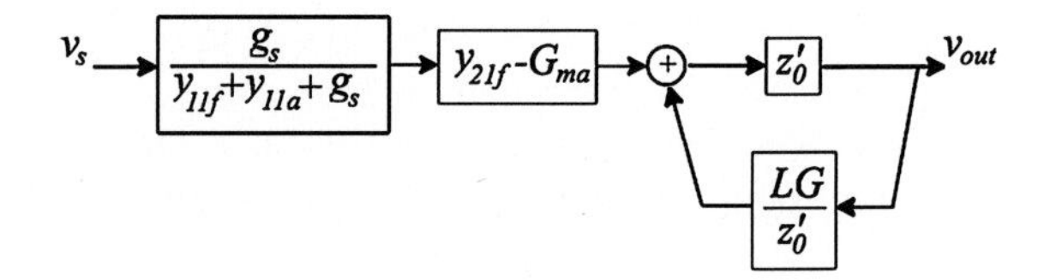

Fig. 5.8 System flow graph substituting for G_{mos} and recognizing z_0' and LG

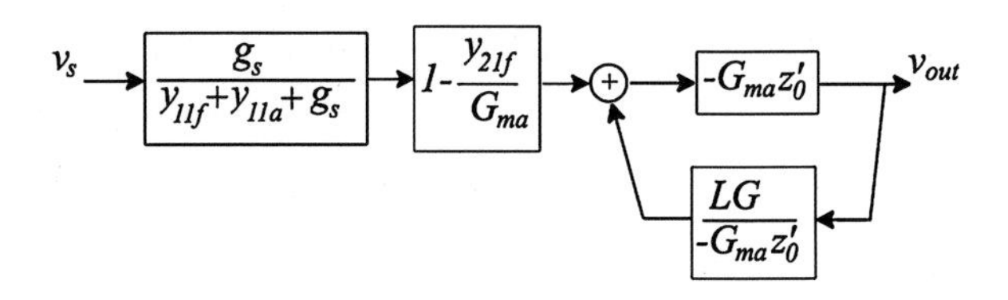

Fig. 5.9 System flow graph with G_{ma} moved into loop

The system voltage transfer function, H_{os}, can now be written directly as Eq. (5.10).

$$
\begin{aligned}
H_{\mathrm{os}} &= g_{\mathrm{s}} \cdot \frac{1}{g_{\mathrm{s}} + y_{11\mathrm{f}} + y_{11\mathrm{a}}} \cdot \left(1 - \frac{y_{21\mathrm{f}}}{G_{\mathrm{ma}}}\right) \cdot \frac{-G_{\mathrm{ma}} z_0'}{1 - \mathrm{LG}} \\
&= -H_{1\mathrm{s,os}} \cdot \left(1 - \frac{y_{21\mathrm{f}}}{G_{\mathrm{ma}}}\right) \frac{A_{\mathrm{OL}}}{1 - LG} \\
A_{\mathrm{OL}} &= -G_{\mathrm{ma}} z_0'
\end{aligned}
\tag{5.10}
$$

This result is significantly more complicated than may have been expected—the voltage-divider factor is a function of source, amplifier, and feedback net, a factor in parenthesis compares feedback with amplifier forward transmission, and the amplifier "open-loop gain," A_{OL}, is seen to be a function of load and feedback net through z_0' and not strictly of the amplifier itself as may be inferred from the tag "*OL*" in this parameter.

This breakdown into components facilitates insight to the design process. The transfer function is composed of factors due to subsets of the design and can be individually examined and modified. Some variables will affect more than one factor since the design is not completely decoupled into these factors. $H_{1s,os}$ is voltage transfer function at the amplifier inverting input due to the source with output shorted to ac ground. It is dependent on source impedance and input impedance parameters of the amplifier and of the feedback net. y_{21f} is the forward transfer transconductance of the feedback net, and G_{ma} is that of the amplifier. The magnitude of the factor in parenthesis is approximately one when the amplifier transconductance dominates and larger than one when the feedback path dominates. The larger of these transconductance parameters indicates if the signal path is dominated by the amplifier or by the feedback net. A_{ol} is a modified "amplifier open-loop gain," explicitly a function of the amplifier's transconductance and the modified impedance at the output node with the feedback zeroed (opened in the graph). *LG* is the system loop gain as we have developed above. It too is in factored form such that design information is evident.

The Single-Transistor Feedback Gain Cell

Consider the single-transistor gain cell in Fig. 5.10a. We can analyze this by inspection for the short circuit, i_{sc}, generating the flow graph in Fig. 5.10b. Using the simple model for the transistor with only g_m and g_{ds}, we write for the short-circuit current, Eq. (5.11).

$$\begin{aligned} i_{\mathrm{sc}} &= g_{\mathrm{s}} \cdot \frac{1}{g_{\mathrm{s}} + g_{\mathrm{f}} + sc_{\mathrm{f}}} \cdot (g_{\mathrm{f}} + sc_{\mathrm{f}} - g_{\mathrm{m}}) \cdot v_{\mathrm{s}} \\ &= \frac{r_{\mathrm{f}}}{r_{\mathrm{f}} + r_{\mathrm{s}}} \cdot \frac{1}{1 + \dfrac{sc_{\mathrm{f}}}{g_{\mathrm{f}} + g_{\mathrm{s}}}} \cdot (-g_{\mathrm{m}}) \cdot \left(1 - \frac{1 + sc_{\mathrm{f}} r_{\mathrm{f}}}{g_{\mathrm{m}} r_{\mathrm{f}}}\right) \cdot v_{\mathrm{s}} \\ &= \frac{-r_{\mathrm{f}} \cdot g_{\mathrm{m}}}{r_{\mathrm{f}} + r_{\mathrm{s}}} \cdot \frac{1}{1 + \dfrac{sc_{\mathrm{f}}}{g_{\mathrm{f}} + g_{\mathrm{s}}}} \cdot \left(1 - \frac{1 + sc_{\mathrm{f}} r_{\mathrm{f}}}{g_{\mathrm{m}} r_{\mathrm{f}}}\right) \cdot v_{\mathrm{s}} \end{aligned} \tag{5.11}$$

z_0' is also found by inspection as the parallel combination of branch impedances at the output node without the "feedback" loop currents due to the amplifier transconductance in Eq. (5.12), ignoring the crosscurrents at the port.

$$\begin{aligned} z_0' &= \frac{1}{g_{\mathrm{load}} + sc_{\mathrm{load}} + g_{\mathrm{ds}} + g_{\mathrm{d}} + \dfrac{1}{r_{\mathrm{s}} + \dfrac{r_{\mathrm{f}}}{1 + sc_{\mathrm{f}} r_{\mathrm{f}}}}} \\ &= \frac{1}{g_{\mathrm{load}} + sc_{\mathrm{load}} + g_{\mathrm{ds}} + g_{\mathrm{d}} + \left(g_{\mathrm{f}} || g_{\mathrm{s}}\right) \cdot \dfrac{1 + sc_{\mathrm{f}} r_{\mathrm{f}}}{1 + sc_{\mathrm{f}}\left(r_{\mathrm{s}} || r_{\mathrm{f}}\right)}} \end{aligned} \tag{5.12}$$

Loop gain can also be written by inspection. From Eq. (5.7) we see that loop gain is the ratio of minus the sum of the crosscurrents to the sum of the self-currents.

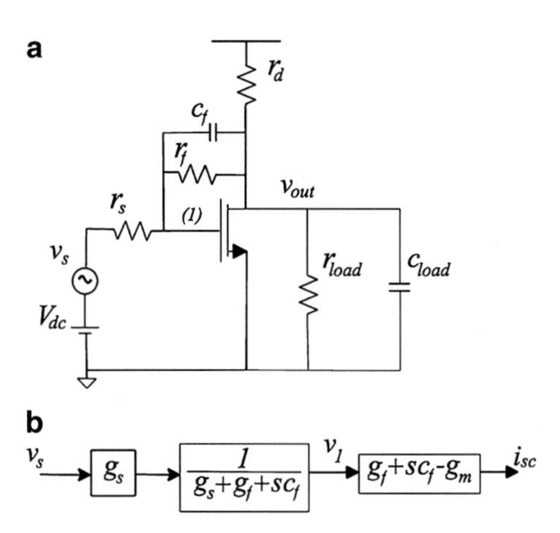

Fig. 5.10 1-Transistor gain cell with feedback (**a**) schematic and (**b**) flow graph for i_{sc}

The reciprocal of the sum of the self-currents is z_0'. There is no reverse transfer in the MOS transistor model used so the only crosscurrent is in the forward direction of the amplifier, and loop gain LG is as shown in Eq. (5.13).

$$\mathrm{LG} = z_0' \cdot \frac{r_\mathrm{s}}{r_\mathrm{s} + \dfrac{r_\mathrm{f}}{1 + r_\mathrm{f} c_\mathrm{f} s}} \cdot (-g_\mathrm{m}) \tag{5.13}$$

Combining the factors we write the full transfer function for the single-transistor feedback gain cell as Eq. (5.14).

$$\begin{aligned} H_\mathrm{os} = & \frac{-r_\mathrm{f} \cdot g_\mathrm{m}}{r_\mathrm{f} + r_\mathrm{s}} \cdot \frac{1}{1 + \dfrac{s c_\mathrm{f}}{g_\mathrm{f} + g_\mathrm{s}}} \cdot \left(1 - \frac{1 + s c_\mathrm{f} r_\mathrm{f}}{g_\mathrm{m} r_\mathrm{f}}\right) \\ & \cdot \frac{1}{g_\mathrm{load} + s c_\mathrm{load} + g_\mathrm{ds} + g_\mathrm{d} + \dfrac{g_\mathrm{f} g_\mathrm{s}}{g_\mathrm{f} + g_\mathrm{s}} \cdot \dfrac{1 + s c_\mathrm{f} r_\mathrm{f}}{1 + s c_\mathrm{f} \left(r_\mathrm{s} \| r_\mathrm{f}\right)}} \\ & \cdot \frac{1}{1 - \dfrac{r_\mathrm{s}}{r_\mathrm{s} + \dfrac{r_\mathrm{f}}{1 + r_\mathrm{f} c_\mathrm{f} s}} \cdot (-g_\mathrm{m}) \cdot z_0'} \cdot v_\mathrm{s} \end{aligned} \tag{5.14}$$

where the top line shows factors due to the effective transconductance G_{mos} of the gain cell, the center line is z_0', and the bottom line is the factor $1/(1-LG)$. The design process is aided by this factored form solution as each factor can be examined for its contribution to the total transfer function.

Examining the net transfer function at low frequency, we set $s=0$ in Eq. (5.14) and, assuming high effective gain $(g_\mathrm{m} z_0')$ at low frequency, we get Eq. (5.15).

$$\begin{aligned} H_\mathrm{os} &= \frac{-r_\mathrm{f} \cdot g_\mathrm{m}}{r_f + r_s} \cdot z_0' \cdot \frac{1}{1 - \dfrac{r_\mathrm{s}}{r_\mathrm{s} + r_\mathrm{f}} \cdot (-g_\mathrm{m}) \cdot z_0'} \\ &= \frac{-r_\mathrm{f}}{r_\mathrm{s}} \end{aligned} \tag{5.15}$$

These results are complete, and we can use this approach to analyze general cells and will do so later in this text. For the present, let us work a bit more with this model and extract some more information. To reach the limiting form in Eq. (5.15), we need to design the cell properly, more than the simple limiting ratio of $-r_f/r_s$. For a gain of *-10*, we start with setting $r_f = 10 \times r_s$.

In Eq. (5.14) we have three primary factors making up the voltage transfer function for the gain cell. The first line is the circuit's short-circuit current. The factor in parenthesis is due to the relative values of the feedback resistor to the transistor's forward transconductance. At low frequency, for $g_m \times r_f = 20$, this factor alone limits the cell gain to -9.5. This fact gives us another design relation, Eq. (5.16).

$$g_m \gg \frac{20}{r_f} \tag{5.16}$$

Using $r_f = 1\ M\Omega$, $r_s = 100\ k\Omega$ for the limiting gain of *10*, a cascoded current source at the drain node, and a capacitive load only, we get the output response as shown in Fig. 5.11 from a closed-loop ac simulation. Low-frequency gain is *19.5 dB* for a linear gain of *9.44* (using values for technology, 3 V, 0.35 μm, nMOS $L = 4$ μm, $W = 10$ μm, $M = 4$, $I_{ds} = 30\ \mu A$).

The three components, i_{sc}, z_0', and loop gain, are also found from simulation and shown in Fig. 5.12 where the top trace is the short-circuit current at the output due to v_s, second trace is z_0', third trace is the system loop gain *LG*, and the bottom trace is the phase plot for *LG*. Note the loop gain phase at low frequency is *180°*.

Substituting the low-frequency values from the plots in Fig. 5.12, we calculate the low-frequency closed-loop gain as *9.46*, Eq. (5.17), in agreement with the closed-loop simulation result in Fig. 5.11. And the reason we are not seeing a gain of *−10* is found in the relatively small loop gain on *18.75* in the denominator. Increasing the loop gain by a factor of 100 would get us to a closed-loop gain of *−9.96*.

$$H_{os}(0) = i_{sc} \cdot \left. \frac{z_0'}{1 - LG} \right|_{s=0} = 570.5\,\mu A \cdot \frac{690.7\,k\Omega}{1 + 18.75} = 9.46 \tag{5.17}$$

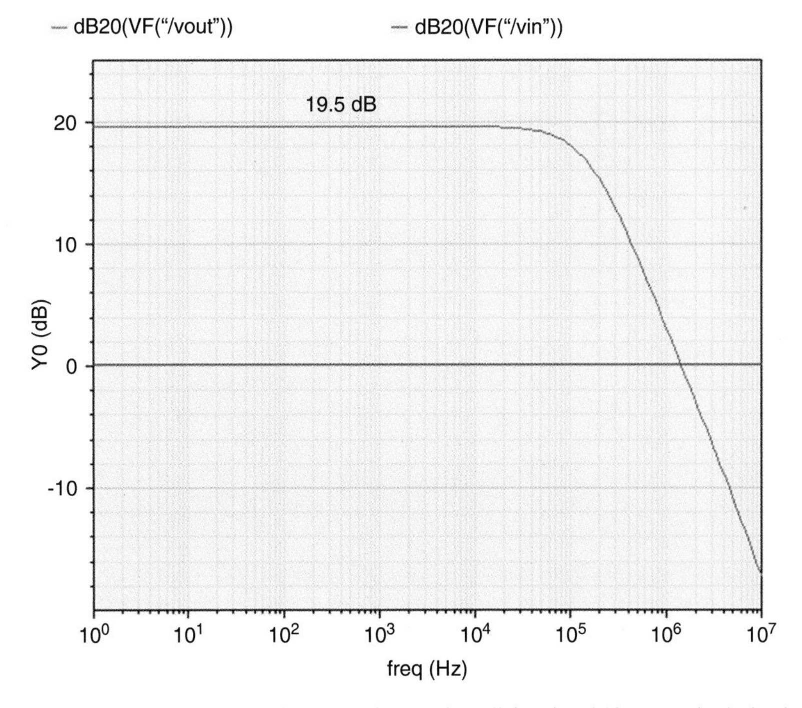

Fig. 5.11 AC response for the single-transistor gain cell in Fig. 5.10a, magnitude in dB vs frequency

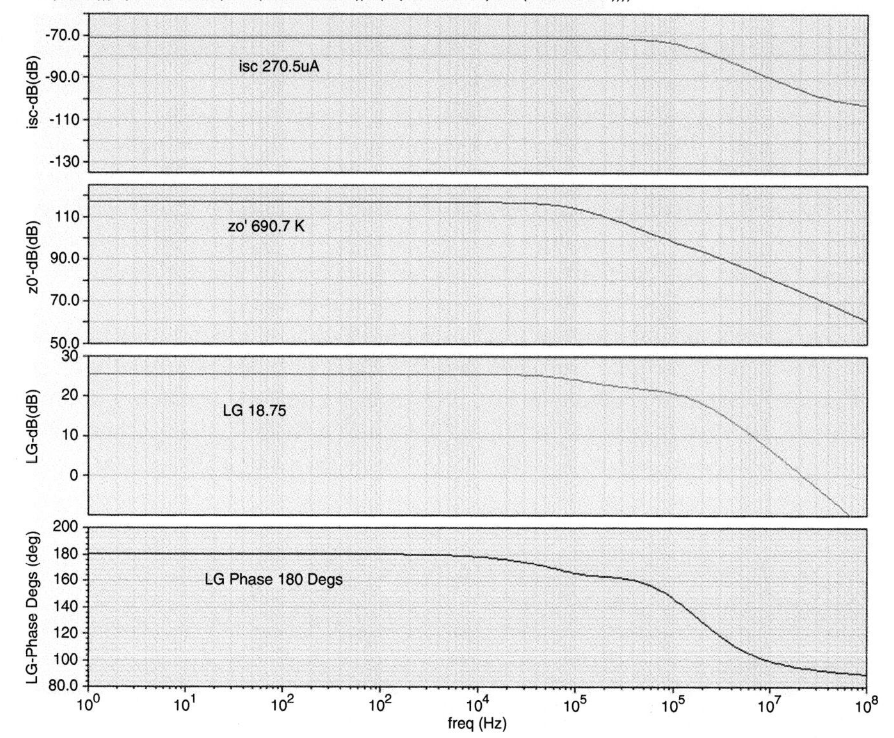

Fig. 5.12 AC response for single-transistor gain cell in Fig. 5.10, top (1) – i_{sc}, (2) z_O', (3) LG, (4) LG phase

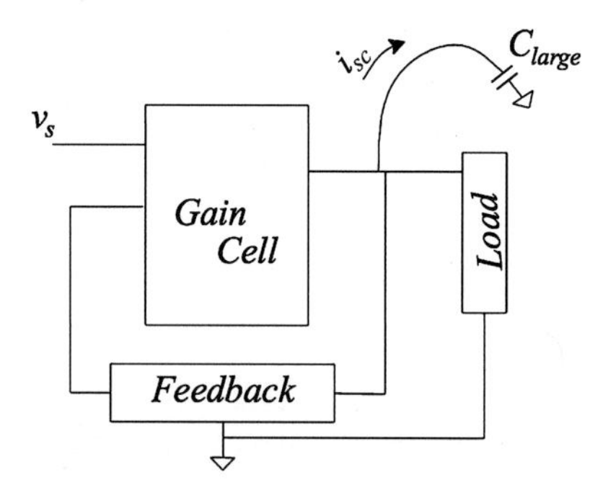

Fig. 5.13 Simulation setup for G_{mos} (=i_{sc} in Fig. 5.12) for the single-transistor gain cell in Fig. 5.10

Simulating Transfer Function Factors

The process we have developed allows for hand analysis explorations for some of the factors (using simple models for active devices). The use of simulators enhances this analysis significantly and allows for the use of full device models and wiring parasitic elements if desired. Figure 5.12 shows some simulation results of transfer function factors, revealing locations and origins for some of the poles and zeros in the net result.

The setup of simulations for particular factors is guided by the flow graph. We can operate at a high level as indicated in Fig. 5.6 and use a single node selected in the outer feedback loop, or we can get more information using additional nodes such as in the flow graph in Fig. 5.10 where the gate node for the nMOS transistor is included. Here we will use the single-node graph to demonstrate some setups.

To obtain G_{mos} we need to find the current into an ac-short at the output node due to v_s, the input excitation. By adding a very large capacitor to the output node, we make v_{out} an ac ground node, disable the feedback loop, and maintain the proper DC voltage at the node. This result for the single-transistor gain cell is shown in the top plot of Fig. 5.12.

To simulate for z_0' and loop gain, LG, we use the simulation arrangement as shown in Fig. 5.14. Two copies of the feedback system are instantiated, each with a break in the outer loop, here at the output node itself. A very large inductor provides ac isolation while DC bias conditions are introduced by the DC voltage source (this can be set from a third instance in closed loop, using a VCVS from an appropriate node). A large capacitor can be used here as well. Each of the instances is shown with two ac sources. Sources v_{11} and v_{22} will have ac magnitude 1 while sources v_{12} and v_{21} and both v_s inputs will be set to zero ac magnitude. Circuit properties z_0' and loop gain LG are obtained by forming the relations in Eq. (5.18) from the simulation results as shown in Fig. 5.12 for the single-transistor gain cell as the 2nd plot from the top for z_0' and as the 3rd plot for loop gain, LG.

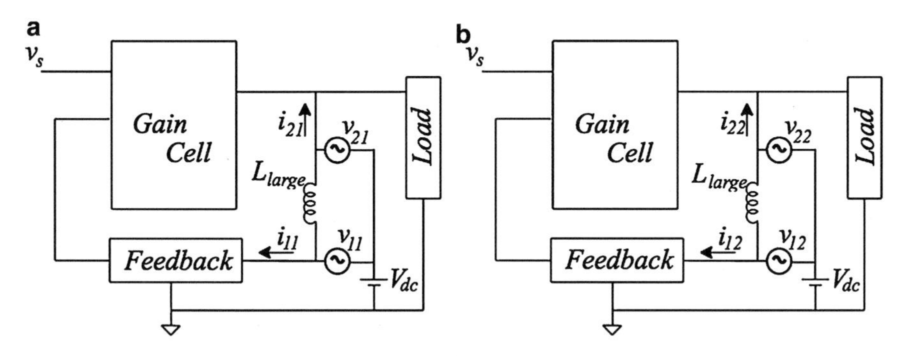

Fig. 5.14 Simulation setup for z_0' and LG

$$z_0' = \frac{1}{i_{11} + i_{22}}$$
$$\mathrm{LG} = \frac{-(i_{12} + i_{21})}{i_{11} + i_{22}} \qquad (5.18)$$

This process can be shown to be fully equivalent to the return ratio method using the forward controlled source of a two-port amplifier representation for designs having a unidirectional amplifier. Amplifiers having reverse transmission will have loop energy in both directions requiring the symmetry shown in Eq. (5.18) that is not included in the Return Ratio analysis. This is particularly important for high-frequency designs where parasitic capacitive couplings provide this reverse transmission through active devices as well as through interconnect interactions.

Using this process we can properly setup the open-loop method generally used. It is obvious that you cannot simply open the loop in a feedback system, drive one end, and monitor the other end of the break. For this approach to approximate the proper result, the break is required to be at a proper location having either high or low impedance to minimize error, *or the loop must be properly terminated.* The impedance at the drive side of the break is not included in the loop gain obtained by simply breaking the loop. This impedance is sometimes included by the use of a replica load—a copy of the cell at the break also in open-loop condition. While this is an improvement, the replica itself must be properly terminated as shown in Fig. 5.15. The loop is opened using a large inductor at a point in the outer loop, here at the amplifier output/feedback net wire shown in the bottom instance of the

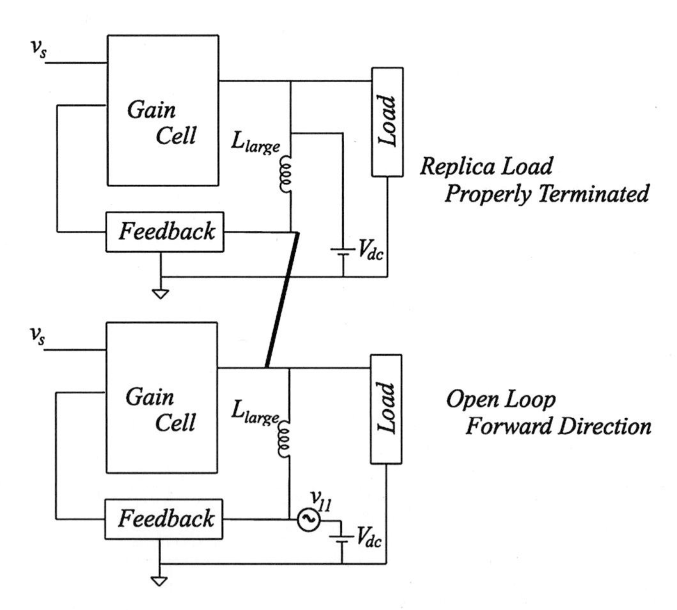

Fig. 5.15 Simulation setup for properly replica loading the open-loop method for finding loop gain

feedback system. DC conditions are imposed at the break, and the feedback net is driven by a unit magnitude ac-source, v_{l1} (with $v_s = 0$). Loop gain is taken at the amplifying cell output at the opposite side of the isolating inductor. Note that the replica maintains the DC condition at the break directly at the output port and through the inductor at the feedback net input side. As with the return ratio method, this correction is only for the forward loop direction. To account for the reverse loop gain portion, we use another pair of instances driven in the reverse direction and add the results to account for reverse as well as forward loop transmission as automatically done in using the Z-Method.

Stability and Phase Margin

While feedback can be beneficial in an analog design, it can also produce unwanted behavior. A system transfer function can be written in a form similar to that in Eq. (5.17). At some frequency $f_{lgunity}$, the loop gain function magnitude will go to unity, and the denominator can approach zero depending on the phase of LG that can be written as $1e^{j\varphi}{}_{lgunity}$ at this frequency, Eq. (5.19). A phase of $0°$ results in the condition of dividing by zero, and the system shows very large gain and may oscillate. Phase margin is the measure of how closely the system comes to this condition of vanishing denominator, the difference between the phase of LG at unity magnitude and the critical phase of zero degrees. (Many definitions use 180° as the critical phase due to the external subtraction in the system feedback model.)

$$\begin{aligned} H_{os}\left(f_{\text{lgunity}}\right) &= i_{sc}\left(f_{\text{lgunity}}\right) \cdot \frac{z_0'\left(f_{\text{lgunity}}\right)}{1 - \text{LG}\left(f_{\text{lgunity}}\right)} \\ &= i_{sc}\left(f_{\text{lgunity}}\right) \cdot \frac{z_0'\left(f_{\text{lgunity}}\right)}{1 - e^{j\varphi_{\text{lgunity}}}} \end{aligned} \quad (5.19)$$

We now apply the analysis method developed for feedback analysis to the operational transconductance amplifier (OTA) gain stage in Fig. 5.16 to find system transfer function and examine its loop gain and phase margin components. This gain stage is an improvement on the single-transistor design we saw in Fig. 5.10.

To find the transfer function for this system, we again write the three relations: i_{sc}, z_0', and LG in Eq. (5.20), where we model the OTA input stage as an ideal transconductance with output admittance $g_{outl} + sc_{outl}$. We do this to keep the algebra manageable while still showing design features.

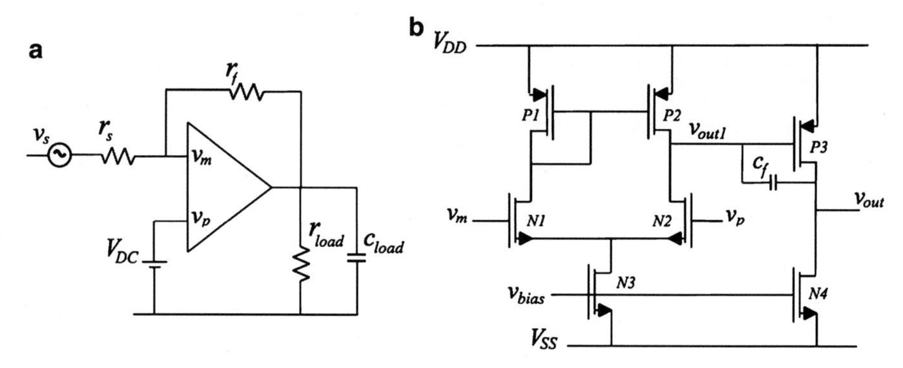

Fig. 5.16 (**a**) Op amp gain stage and (**b**) two stage OTA schematic

$$i_{\mathrm{sc}} = g_{\mathrm{s}} \cdot \frac{1}{g_{\mathrm{f}} + g_{\mathrm{s}} + sc_{\mathrm{i}}} \cdot \left(g_{\mathrm{f}} + \frac{g_{\mathrm{mn}} \cdot \left(sc_{\mathrm{f}} - g_{\mathrm{mp3}} \right)}{g_{\mathrm{out1}} + sc_{\mathrm{out1}} + sc_{\mathrm{f}}} \right) \cdot v_{\mathrm{s}}$$

$$z_0' = \frac{1}{g_{\mathrm{ld}} + sc_{\mathrm{ld}} + g_{\mathrm{out}} + \dfrac{1}{r_{\mathrm{f}} + \dfrac{1}{g_{\mathrm{s}} + sc_{\mathrm{i}}}} + \dfrac{1 + \dfrac{g_{\mathrm{mp3}}}{sc_{\mathrm{out1}} + g_{\mathrm{out1}}}}{\dfrac{1}{sc_{\mathrm{f}}} + \dfrac{1}{g_{\mathrm{out1}} + sc_{\mathrm{out1}}}}} \tag{5.20}$$

$$\mathrm{LG} = z_0' \cdot \frac{g_{\mathrm{s}}}{g_{\mathrm{s}} + sc_{\mathrm{i}}} \cdot \frac{g_{\mathrm{mn}} \cdot \left(sc_{\mathrm{f}} - g_{\mathrm{mp3}} \right)}{g_{\mathrm{out1}} + sc_{\mathrm{out1}} + sc_{\mathrm{f}}}$$

We combine these relations as in Eq. (5.19) to generate the input-output transfer function. The system loop gain, *LG*, is seen to have two poles plus a right-half plane zero in addition to the poles and zeros of z_0', with the pole/zero configuration for z_0' not as clear. To get a better idea of these, we resort to simulation, first using an ideal transconductor for the input differential OTA stage and the following parameters:

r_f	5 MΩ	
r_s	500 kΩ	
g_{mn}	64 μMohs	Ideal input differential stage transconductance
g_{mp3}	81 μMohs	From Simulation
g_{dsp3}	48 nMohs	From Simulation
g_{dsn4}	1648 nMohs	From Simulation
g_{out1}	100 nMohs	
c_{out1}	500 fF	Capacitance due to output pMOS
c_f	3 pF	Internal compensation capacitor
r_{load}	10 MΩ	
c_{load}	2 pF	

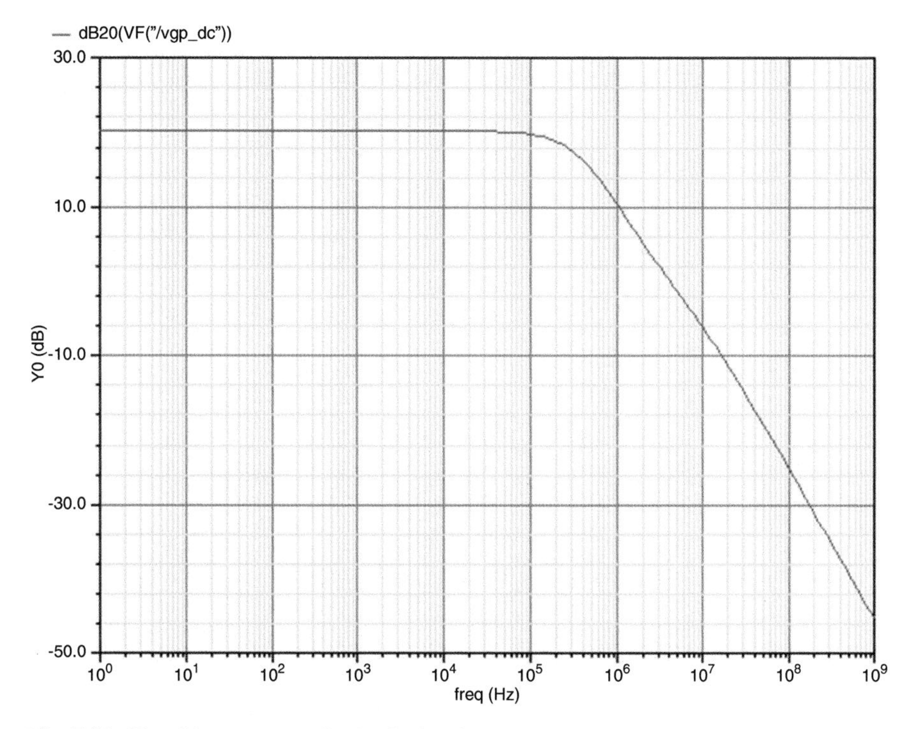

Fig. 5.17 Closed-loop response for the OTA gain stage shown in Fig. 5.16

The closed-loop response as shown in Fig. 5.17 has a *20 dB* low-frequency gain and a *−3 dB* bandwidth of *350 kHz*. No peaking is seen indicating good phase margin.

Figure 5.18 shows the currents i_{11} and i_{22}, top plot, and the phase for i_{22} in the bottom plot. The feedback path current, i_{11}, is flat with frequency and always less than that into the load and OTA, i_{22}. The dominate i_{22} current shows a zero at *~19 Hz*, a pole at *4.3 kHz* and another zero at *~6.9 MHz*. The phase plot below shows the zeros to be in the left-half plane. The reciprocal of the sum of these currents is the port impedance with feedback zeroed, z_0' in Eq. (5.20), shown in Fig. 5.19. The internal feedback loop is included in this result.

The numerator in the system transfer function is the product of z_0' and the output node short-circuit current. i_{sc} can be obtained using a setup as shown in Fig. 5.13 and Fig. 5.20. Here we see the effect of the right-half plane zero with the phase decreasing another 90°.

Figure 5.21 shows two plots. The upper plot shows four traces—in order of larger low-frequency magnitude—these are (1) transfer function numerator, (2) loop gain, (3) closed-loop transfer function, and (4) a *0 dB* reference signal. While the loop gain function is greater than one, the output transfer function is the difference in dB's between these two traces, about 20 dB shown as the third trace. Once the loop gain function falls below unity at about *100 kHz*, the transfer function

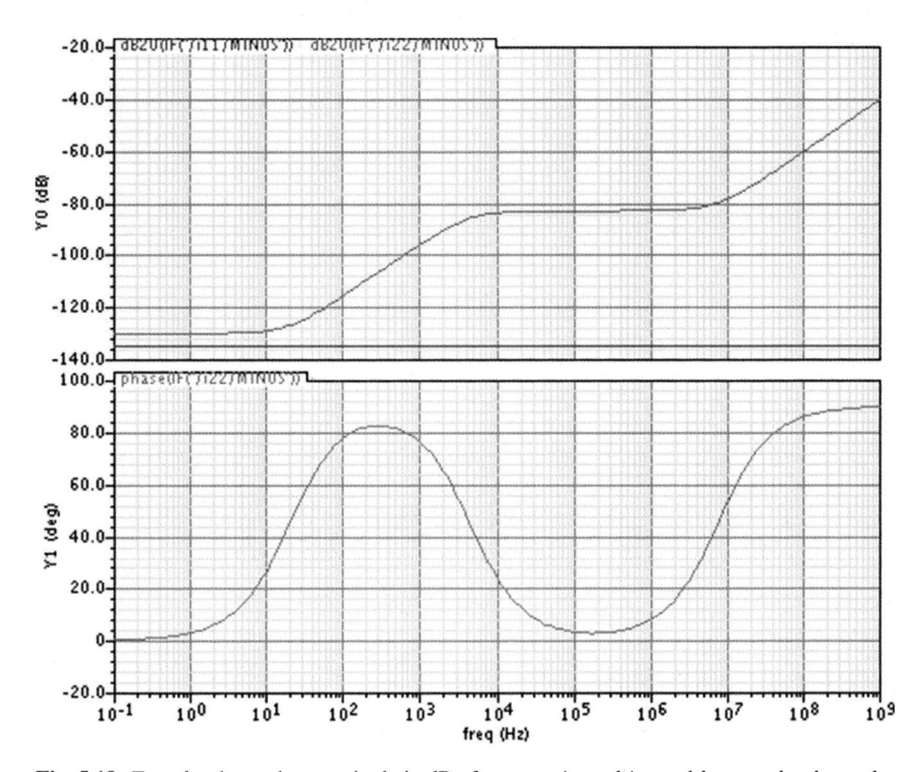

Fig. 5.18 *Top plot* shows the magnitude in dB of currents i_{11} and i_{22}, and *bottom plot* shows the phase for the dominate z_0' current, i_{22}, versus frequency

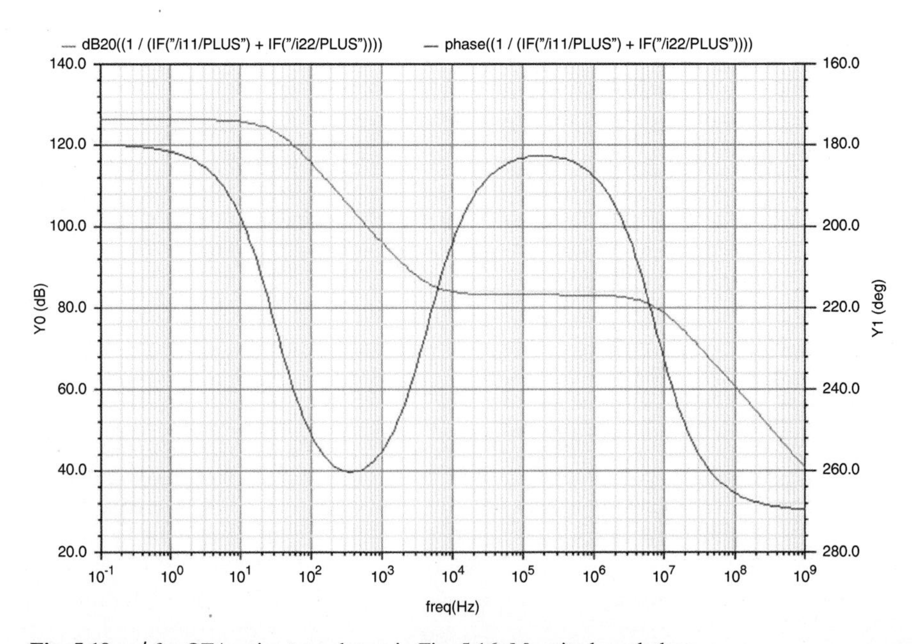

Fig. 5.19 z_0' for OTA gain stage shown in Fig. 5.16. Magnitude and phase

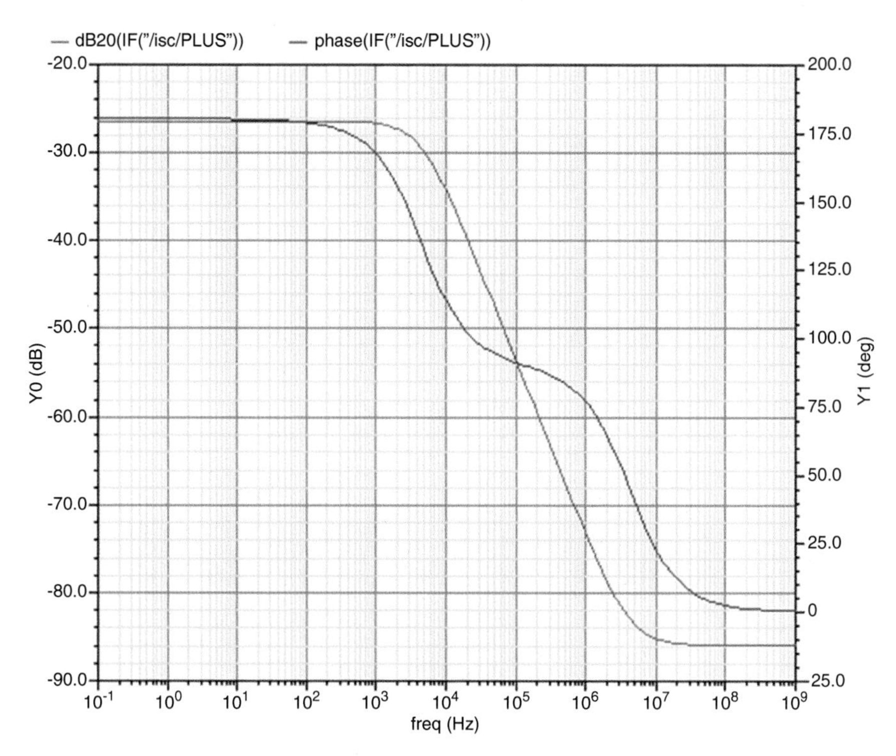

Fig. 5.20 i_{sc} for OTA gain stage shown in Fig. 5.16. Magnitude and phase

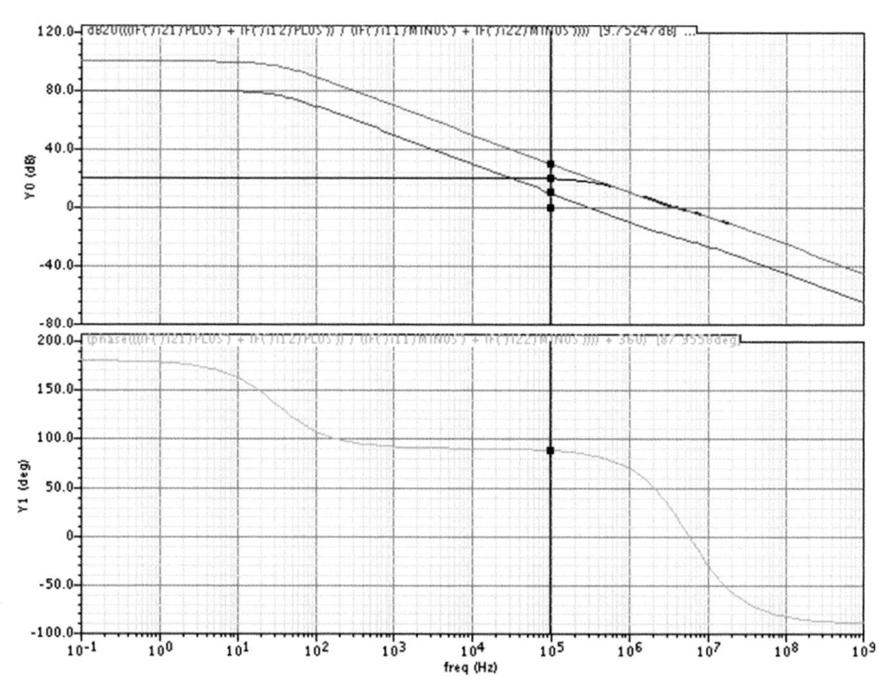

Fig. 5.21 *Top*: transfer function numerator, loop gain, and closed-loop response. *Bottom*: loop gain phase for OTA gain stage shown in Fig. 5.16 with ideal input differential stage

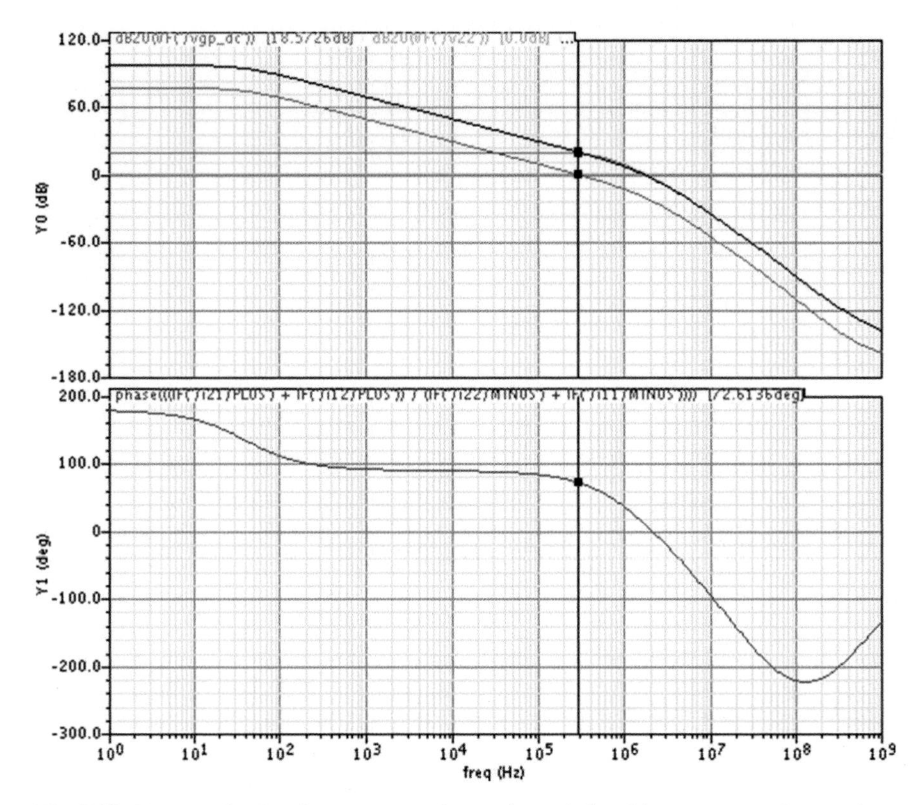

Fig. 5.22 *Top*: transfer function numerator, loop gain, and closed-loop response. *Bottom*: loop gain phase for OTA gain stage shown in Fig. 5.16 with transistor level differential stage

denominator goes to one, and the transfer function tracks the numerator. The lower plot is the loop gain phase showing a phase margin of *83*° for the closed-loop system.

Using a full transistor level schematic, we get the simulation set shown in Fig. 5.22 showing agreement with the simplified results above. At higher frequencies we see deviations caused by the input capacitance as well as response from the input differential stage. Phase margin here is ~*73*° consistent with the closed-loop ac response. Additional simulations can be done to fully account for these responses, and with the decomposition afforded with this approach, modifications can be made and their effects verified easily.

A Unity Gain Buffer

A unity gain buffer is one with full feedback to the inverting input, driven at the non-inverting input, Fig. 5.23a. Following ideal 0-order analysis, this configuration will drive the inverting input and therefore the output, to follow the non-inverting input signal.

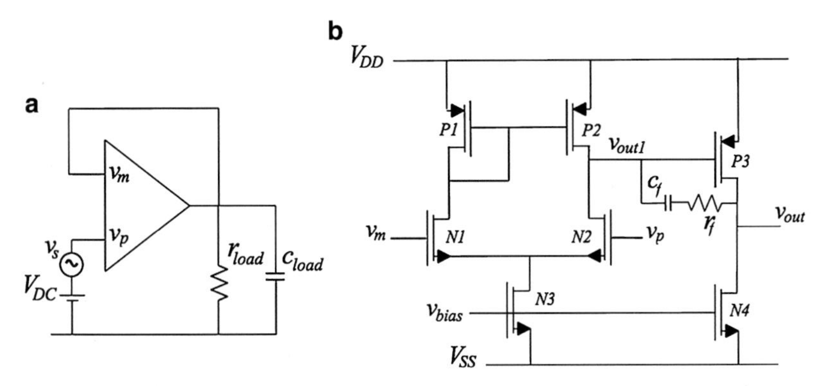

Fig. 5.23 (**a**) Unity gain buffer and (**b**) schematic for OTA used in unity gain buffer

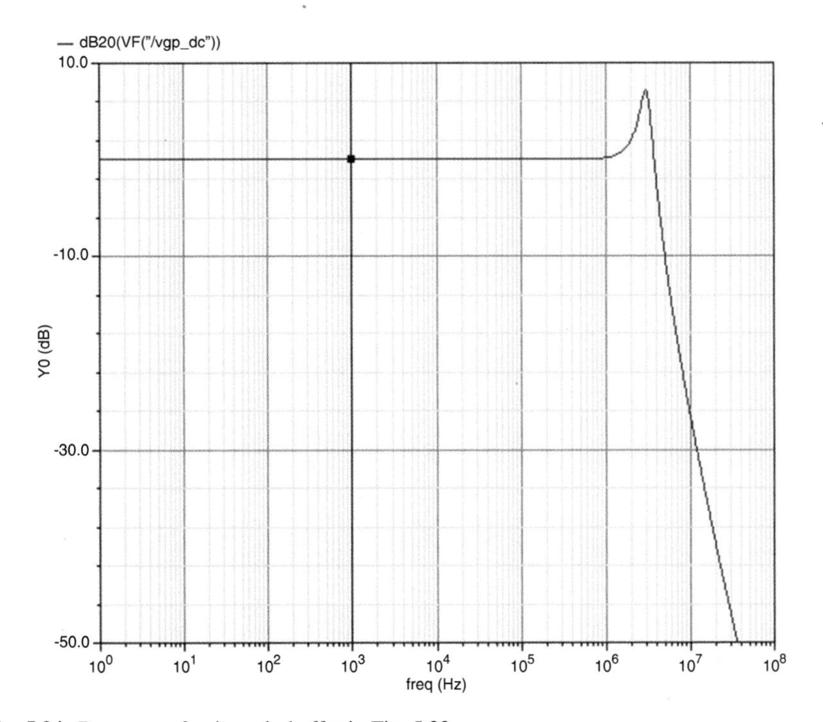

Fig. 5.24 Response of unity gain buffer in Fig. 5.23

The OTA used in this example is shown in Fig. 5.23b. It is the same design as was used in the OTA gain stage with the addition of the resistor r_f in series with the compensation capacitor c_f in the internal feedback loop. This internal loop is automatically included in the calculation of z_0'.

Using the same bias conditions as with the gain stage design in the previous section and with $r_f = 0$, the closed-loop response shows about $7\ dB$ of peaking, Fig. 5.24. The loop gain for this design is shown in Fig. 5.25. Phase margin is about $17°$.

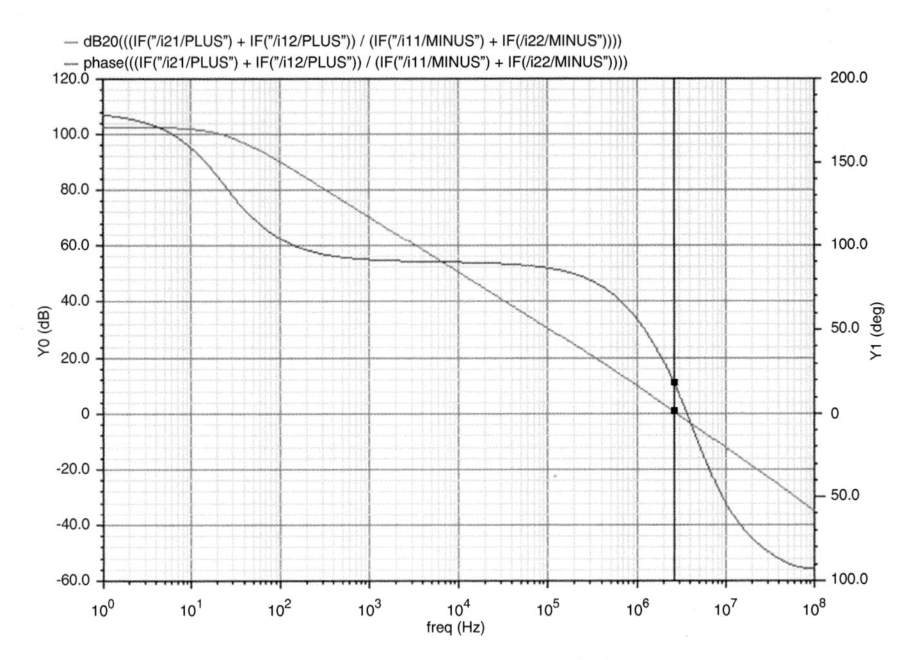

Fig. 5.25 Loop gain magnitude and phase for unity gain buffer in Fig. 5.23

Figure 5.25 shows a single-pole roll-off for loop gain magnitude. This should result in a phase change of $-90°$ and not the $-270°$ change seen, so something else is happening. To explain this and more importantly to correct it, we examine the components for loop gain—the numerator and denominator separately. The reciprocal of the loop gain denominator, z_0', shown in Fig. 5.26, indicates a pole/zero/pole behavior, all in the left-half plane.

The numerator response for the loop gain function, Fig. 5.27, shows a pole followed by a zero and excess phase change indicative of a right-half plane zero. This zero is at about 10^7 Hz, close to a pole location in z_0', Fig. 5.26. We have a pole/zero magnitude cancelation, but with the zero in the right-half plane, the phase of the loop gain function changes by $-180°$, causing the reduction of phase margin.

The RHZ comes from the factor in the crosscurrent i_{21}, Eq. (5.21), due to the series capacitor-resistor from the gate to the drain of the output pMOS transistor, P3. With $r_z = 0$, the zero is at $s = g_{mp}/c_f$. Adding the series resistor moves the zero to higher positive frequencies until $r_z = 1/g_{mp}$ at which point it is located at infinity. Increasing it further moves the zero into the left-half plane. Making $r_z 12\ k\Omega$ with $g_{mp} \sim 81$ µMohs moves the zero sufficiently to higher frequencies that we get a phase margin of ~$49.5°$ as shown in the upper plot in Fig. 5.28 and a closed-loop response without peaking as shown in the lower plot. This change in zero location is used in this type of design to reduce the phase change caused by the compensation capacitor alone.

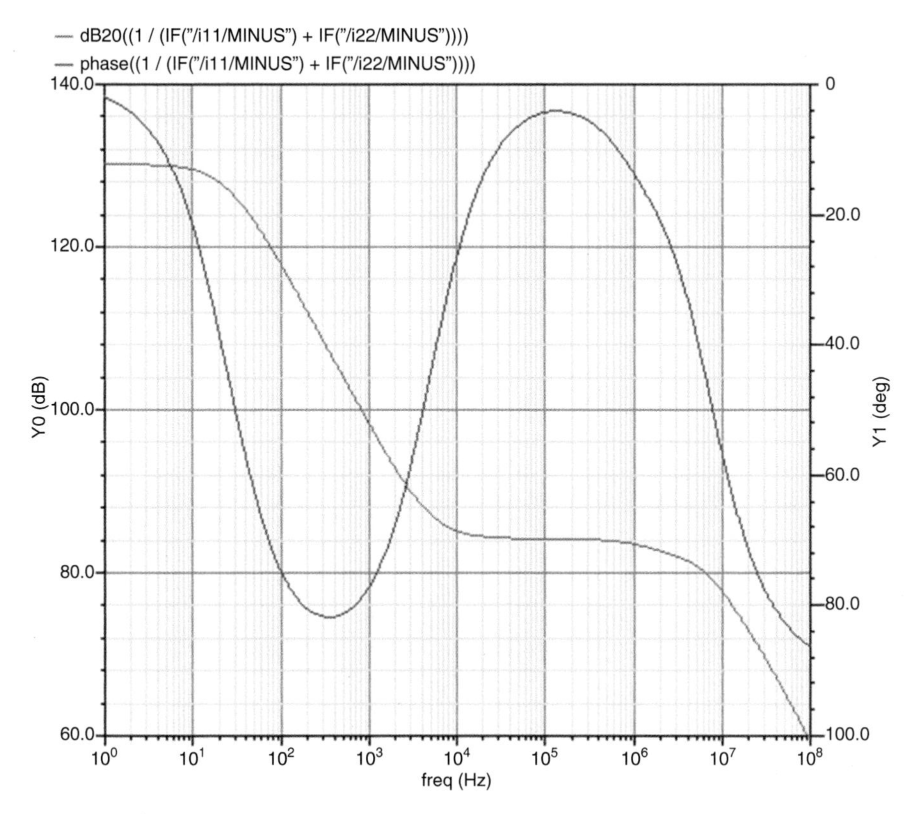

Fig. 5.26 z_0', the reciprocal of the denominator of the loop gain function of the unity gain buffer design, magnitude, and phase

$$
\begin{aligned}
i_{21} &= -g_{\text{m,diff}} \cdot \frac{1}{g_{01} + sc_{\text{i}} + \dfrac{sc_{\text{f}}}{1 + sc_{\text{f}} r_z}} \cdot \left(-g_{\text{mp}} + \frac{sc_{\text{f}}}{1 + sc_{\text{f}} r_z} \right) \cdot v_{11} \\
&= -g_{\text{m,diff}} \cdot \frac{-g_{\text{mp}}}{g_{01} + sc_{\text{i}} + \dfrac{sc_{\text{f}}}{1 + sc_{\text{f}} r_z}} \cdot \left(\frac{1 - \dfrac{sc_{\text{f}}}{g_{\text{mp}}} \left(1 - g_{\text{mp}} r_z \right)}{1 + sc_{\text{f}} r_z} \right) \cdot v_{11}
\end{aligned} \tag{5.21}
$$

The Crystal Oscillator

Oscillators, used to provide system timing or clocks and other functions, are key blocks in both analog and digital systems. A tuned circuit properly placed in the feedback path of an amplifier can produce oscillation at a design frequency. The crystal oscillator uses a piezoelectric crystal as the resonator and due to its extremely high-quality factor Q, provides for very precise frequency selection and low phase noise. The crystal itself physically deflects in an electric field and,

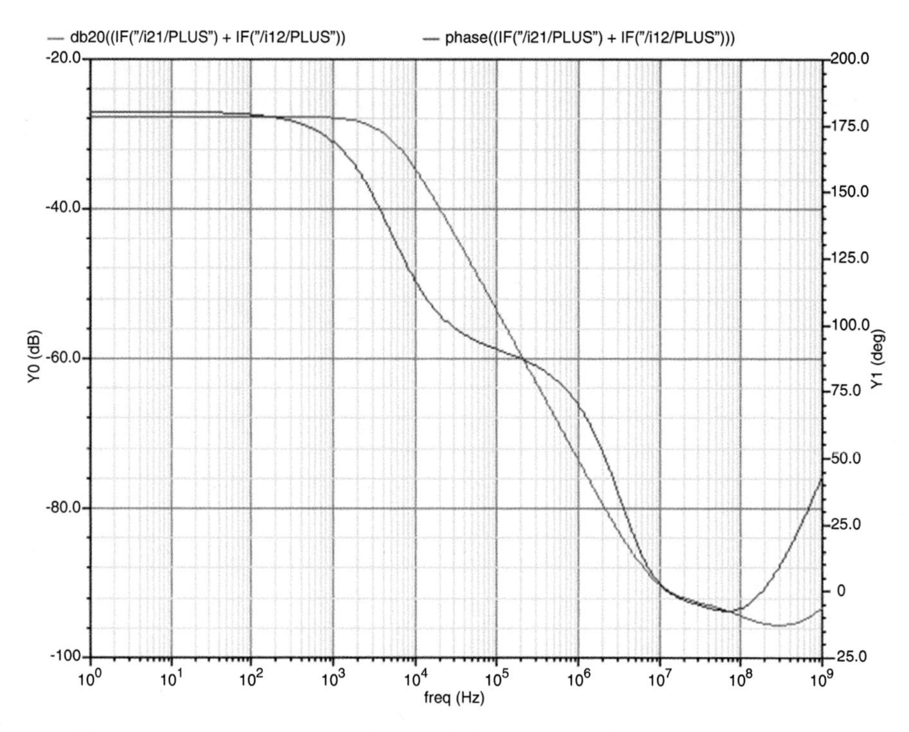

Fig. 5.27 Loop gain numerator, magnitude, and phase showing RHP zero

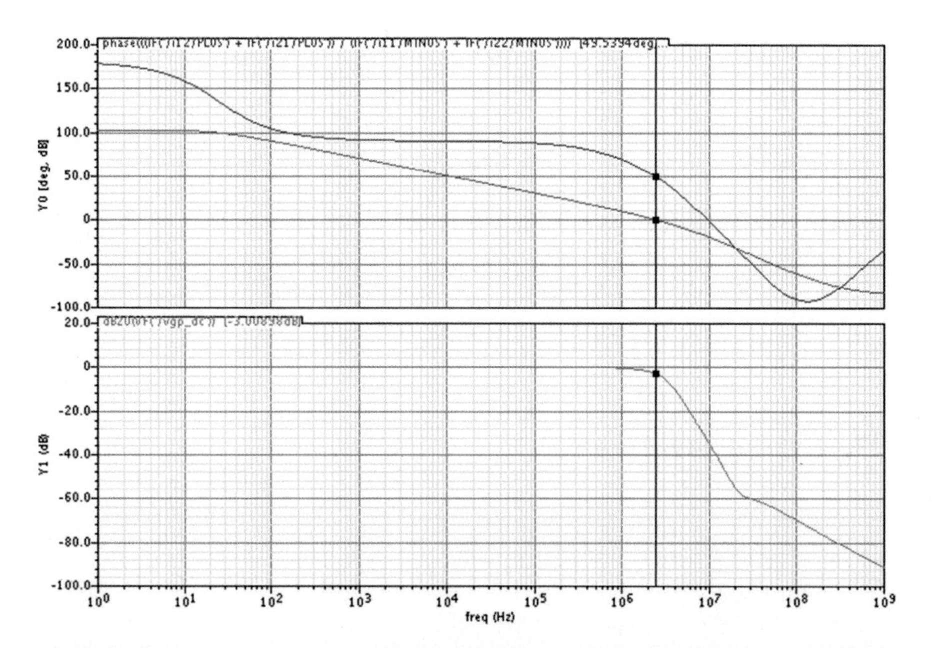

Fig. 5.28 *Bottom plot*: Improved closed-loop response obtained by adding r_z to the OTA compensation and *top plot*—modified loop gain magnitude and phase showing *49.5°* of phase margin

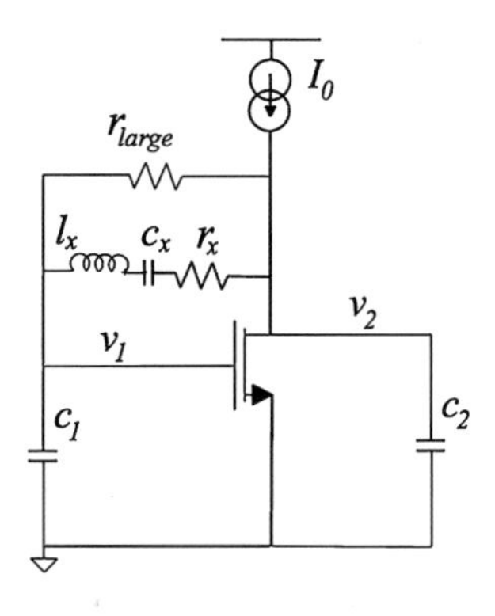

Fig. 5.29 Crystal oscillator circuit—crystal is represented as the series r_x-c_x-l_x branch

when the field is removed, generates a field as the crystal returns to its unstressed shape generating a voltage pulse. The specific shape of the crystal defines the resonant response frequency of a crystal and can be set very accurately. The crystal behaves as an electrical *r-l-c* resonator and as such will not by itself sustain oscillation due to the "resistor" losses so a means of restoring the losses is needed. We need to add energy synchronously each cycle. Consider the circuit in Fig. 5.29, showing a single-transistor amplifier, current source load biasing, and a crystal represented as the r_x-l_x-c_x branch (here we ignore the shunt package capacitance c_0). Resistor r_{large}, usually between *1* and *10 MΩ*, biases the gate at its operating point, and capacitors c_1 and c_2 provide a return path for the resonant crystal current and add phase to the loop propagation.

The crystal manufacturer will specify the net capacitance across the crystal required to achieve the stated operating frequency. This capacitance consists primarily of the series combination of capacitances c_1 and c_2 and is called the "load capacitance" required to move or pull the resonant frequency to the specified operating frequency, normally only by a few ppm. As the crystal is an "off-chip" element, pads must be added to nodes v_1 and v_2 along with ESD protection circuitry adding effective parasitic capacitance onto these nodes and should be accounted for in the total load capacitance.

To find the loop gain, we split the output node v_2 such that c_2 remains with the feedback branch driven by sources v_{11} or by v_{12} in the dual instance simulation bench, and the transistor drain driven by sources v_{22} or by v_{21} in our loop gain formulation. Fig. 5.30 shows the v_{11}/v_{21} part of the loop gain bench. This particular division is not required—c_2 could be grouped with the transistor or we could have selected v_1 as the splitting node. If we assume minimal reverse transmission through the transistor drain to gate, loop gain becomes as shown in Eq. (5.22) where we ignore the reverse loop transmission term.

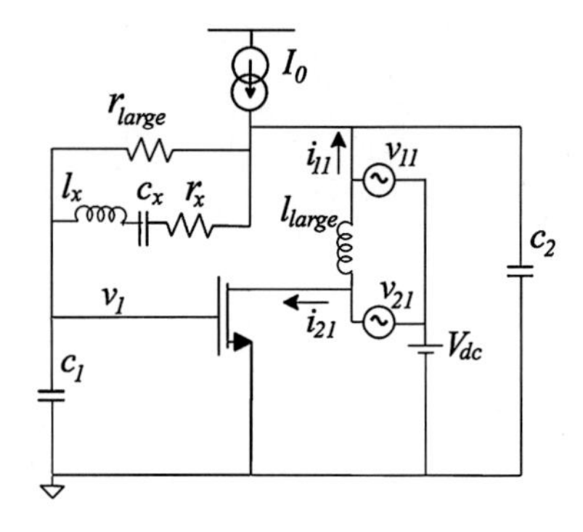

Fig. 5.30 One instance in the loop gain simulation bench showing the setup for crosscurrent i_{21} and self-current i_{11}

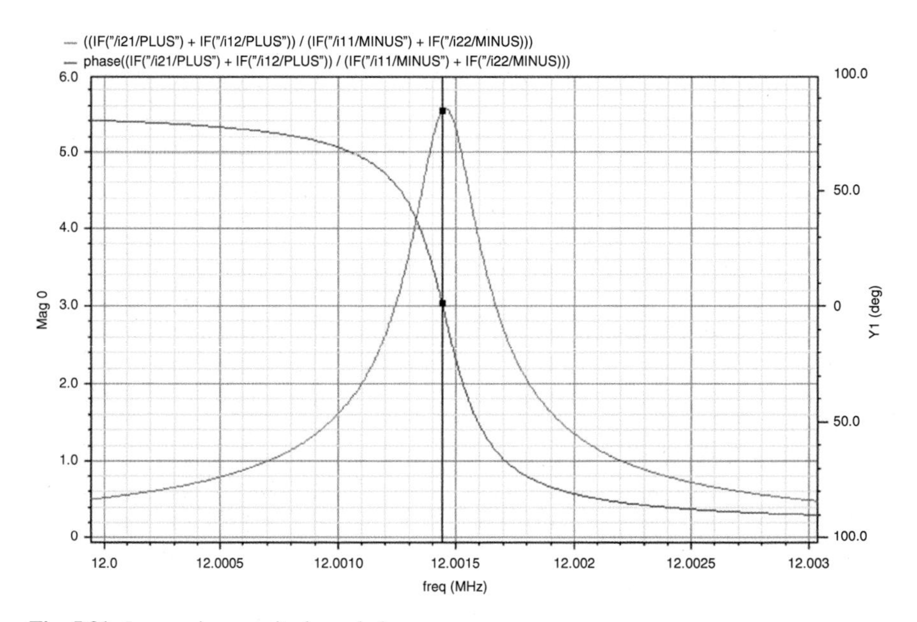

Fig. 5.31 Loop gain, magnitude, and phase

$$
\begin{aligned}
\mathrm{LG} &= -i_{21}.z_0^{'} \\
i_{21} &= \frac{\frac{1}{sc_1}}{\frac{1}{sc_1}+z_x}\cdot(g_{\mathrm{m}})\cdot v_{11} \\
z_0^{'} &= \frac{1}{sc_2+\frac{1}{\frac{1}{sc_1}+z_x}} \\
z_x &= sl_x+\tfrac{1}{sc x}+r_x
\end{aligned}
\tag{5.22}
$$

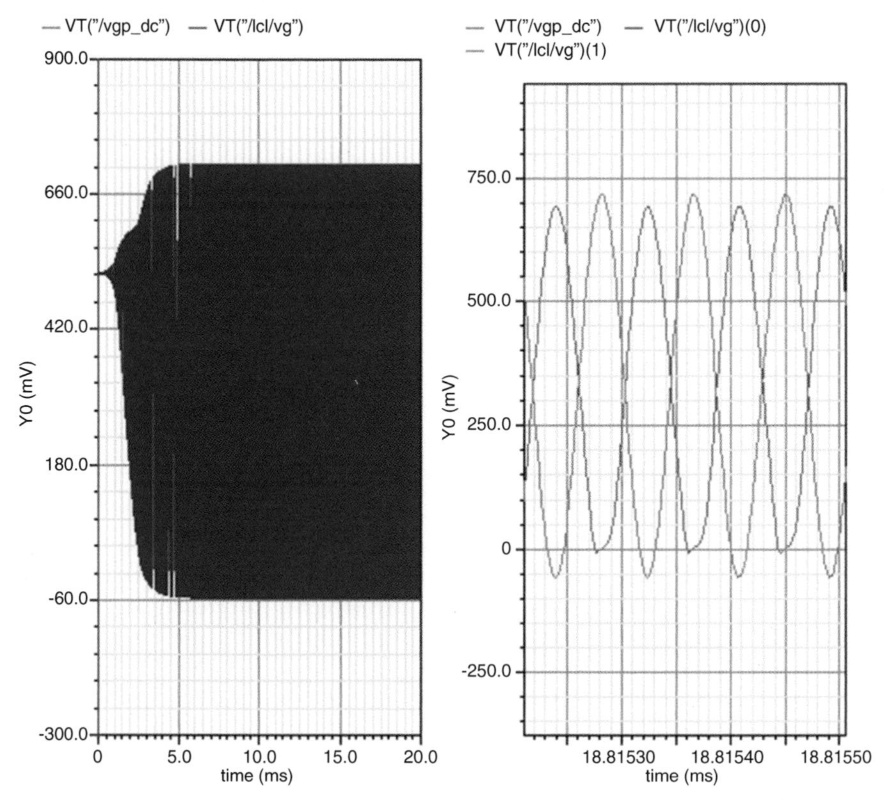

Fig. 5.32 Transient simulation of crystal oscillator circuit (*left*) startup and (*right*) zoom in near steady state, showing response at transistor drain (distorted signal near 0) and at gate

The system loop gain, magnitude, and phase are shown in Fig. 5.31. Phase is seen to change quickly from plus 90° to -90° over a very small change in frequency, crossing zero, for this design, near the peak of the loop gain magnitude of about 5.6. The circuit oscillates as shown in Fig. 5.32 where the left plot shows the startup period and right side shows an expanded region near steady state for the gate and the drain node responses. The drain node shows clipping at the low end. This distortion is the means for amplitude regulation for this circuit. With a loop gain greater than one at zero phase, the system has complex conjugate right-half plane poles, and the response grows exponentially as seen in the early part of the startup transient. As the drain signal increases and becomes distorted due to signal saturation at the low end, the effective loop gain decreases until this effective gain reaches unity, and a balance is reached with the poles now on the $j\omega$ axis.

Summary

This chapter has developed the analysis of feedback circuits using the "z-method," a loop decomposition method that uses driving point impedance and signal flow graphs. This approach differentiates between signals that propagate around the loop from those that go to system ground. Loop Gain derived in Eq. (5.8) and repeated here as Eq. (5.23) is found to be given as the *negative sum of the cross-currents divided by the sum of the self-currents* extracted at a node in the outer loop. This method for loop gain can be performed any place in the outer loop and fully accounts for feedforward, loading, and reverse transmission effects.

$$\mathrm{LG} = \frac{-(i_{21} + i_{12})}{i_{11} + i_{22}} \tag{5.23}$$

The Z-Method is an extension of the return ratio loop gain definition effectively using two-port equivalent modeling for the system at a node in the feedback loop so that it is not limited to using a controlled source in a single active device. In addition, reverse loop transmission is automatically included accounting for bi-directionality of active devices and of active cells. The result in Eq. (5.23) shows the signal symmetry required for the transmission of signals in both directions around the loop.

With this tool, the transfer relations obtained are in the form of Black's classic feedback relation properly identifying his functions of A_{ol} and LG. The amplifier's open-loop gain is shown to be dependent on not only the load and feedback net impedances but also of the input source impedance, Eq. (5.10), making the A_{ol} a misnomer and a source of much misunderstanding of the components of feedback analysis. Another misnomer is the tag "open loop" itself in this term. We do not "open the loop" for this parameter but rather "zero" the loop gain. This operation occurs in the flow graph math domain where the loop is indeed opened but where there is no perturbation on the circuit impedance by this action.

The z-method was applied to a single-transistor gain stage revealing the right-half plane zero and the Miller capacitance multiplier, generalized to arbitrary cell analysis and demonstrated using a 2-stage OTA feedback gain cell. From the flow graphs generated for the analysis of a circuit or system, we derive simulation benches from which we extract particular nodal responses useful in the design. The method reveals behavior issues that can then be isolated to particular nodal voltage and current responses for modifications. This analysis methodology was then used to demonstrate compensation of a unity gain buffer and in the analysis and design of a crystal oscillator.

References

1. H.S. Black, *Stabilized Feedback Amplifiers* (Electrical Engineering, HJan, 1934)
2. P.R. Gray, R. Meyer, *Analysis and Design of Analog Integrated Circuits* (John Wiley and Sons, New York, NY, 1977)
3. A. Ochoa, Analyzing Feedback: Properly Simulating the Open Loop, Mid West Symposium on Circuits and Systems (1998)
4. H.W. Bode, *Network Analysis and Feedback Amplifier Design* (Van Nostrand, New York, NY, 1945)
5. R.D. Middlebrook, Measurement of loop gain in feedback systems. Int. J. Electron. **38**(4), 485–512 (1975)
6. A. Ochoa, Loop gain in feedback circuits: a unified theory using driving point impedance. Mid West Symposium on Circuits and Systems (2013)

Chapter 6
Applications: A Variety of Circuits

Abstract The method of driving point impedance/signal flow graphs for the analysis of analog circuits has been fully developed and extended to the specific analysis of the loop gain function for feedback circuits. In this chapter, we analyze a variety of circuits to fully demonstrate the application of the methodology. In addition, we extend the method to include examples of cells that transform the signal from one representation to another—the phase-locked loop and the switching regulators have cells that take a signal in one domain and output in another—phase to time in the pll, continuous voltage to pulse-width in the switching regulator. Most of these applications contain feedback, giving ample opportunity to analyze feedback effects in circuits with different topologies.

Introduction

The objective of this text is to present a general analysis tool for linear electronic circuits based on the methods of driving point impedance and signal flow graphs. We have already looked at a number of circuits to demonstrate and develop the tool. In this chapter, we look at a variety of additional circuits applying the method to demonstrate it further. The flow graphs are generated mostly by inspection and without the use of equivalent circuits. This is possible as the process works on smaller portions of the circuit at a time, and these partial results are mapped onto the flow graph where they are systematically tied together.

Current Mirrors

Current mirrors are used to copy a reference current and to flip the direction at the delivery node as well as to process a signal by transforming its impedance level or by scaling it. In its simplest realization, two devices are used as shown in Fig. 6.1. This circuit is usually used in a large-signal sense. Large-signal transistor parameters G_M and G_{DS} are defined as shown in Eq. (6.1) where larger differences are implied by using Δ, the difference operator, instead of the differential operator d.

A. Ochoa, *Feedback in Analog Circuits*, DOI 10.1007/978-3-319-26252-9_6

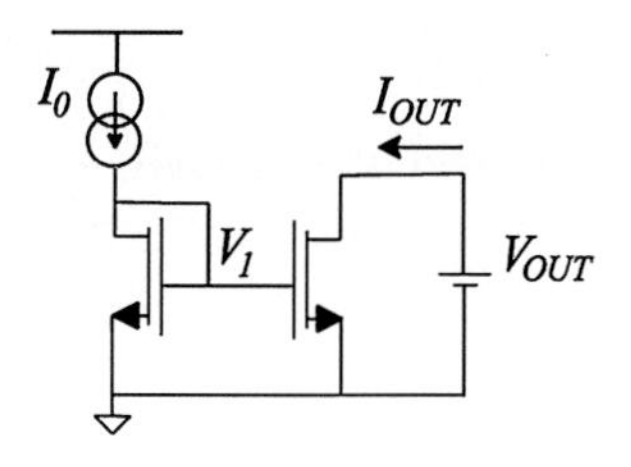

Fig. 6.1 Large-signal current mirror

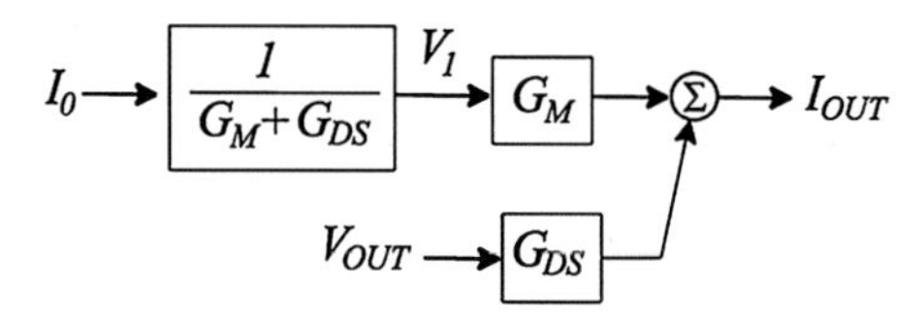

Fig. 6.2 Signal flow graph for large-signal 2 transistor current mirror

Analysis follows small signal algebra, so we will generally use small signal language after this one example.

$$\begin{aligned} G_{\mathrm{M}} &= \frac{\Delta I_{\mathrm{DS}}}{\Delta V_{\mathrm{GS}}} \\ G_{\mathrm{DS}} &= \frac{\Delta I_{\mathrm{DS}}}{\Delta V_{\mathrm{DS}}} \end{aligned} \tag{6.1}$$

We have two nodes in this circuit, making the signal flow graph as in Fig. 6.2. I_0 produces a voltage at node V_1 necessary to absorb that current, part of which flows in the shunt output conductance G_{DS} of the diode connected transistor, and the remaining current flows in the transistor channel as transconductance current. Another transistor having the same geometry and construction (G_M) will produce the same drain channel current. The output transconductor is shunted by the output transistor's large-signal output admittance G_{DS} which will draw current adding to the output current I_{out}. If the output voltage matches V_1, this shunt current matches that in the input branch and I_{out} matches I_0.

$$I_{\mathrm{OUT}} = \frac{I_0}{G_{\mathrm{M}} + G_{\mathrm{DS}}} G_{\mathrm{M}} + G_{\mathrm{DS}} V_{\mathrm{OUT}} \tag{6.2}$$

If we add a small signal current, i_0 to I_0, by superposition, we add the responses. The small signal flow graph is the same as that for the large signal with the respective parameters replaced by g_m and g_{ds}, Eq. (6.3). Because of this parallelism and since we are mostly interested in small signal response, the remaining chapter will assume small signal conditions.

$$i_{\mathrm{out}} = \frac{i_0}{g_{\mathrm{m}} + g_{\mathrm{ds}}} \cdot g_{\mathrm{m}} + g_{\mathrm{ds}} \cdot v_{\mathrm{out}} \tag{6.3}$$

Cascoding the Output Device

The mirroring action error due to the circuit itself (assuming perfect device matching) is due to the output voltage of the mirror transistor compared to that of the control or "diode-connected" transistor. We see in these relations that not all the input current goes into the diode channel and that only the diode channel current is mirrored. The shunt current is compensated at the output to the extent that the output voltage matches the generating diode voltage V_1 for large signal, v_1 for small. We can correct this voltage difference by using the cascode configuration, Fig. 6.3. The DC cascode bias V_{CAS} is chosen such that transistors n_3 and n_4 operate in saturation, making their drain voltages equal, again ignoring mismatch. This modification removes the effect of v_{out} on the mirrored current. It also has the beneficial effect of increasing the output impedance.

Figure 6.4a shows a cascode stage. V_{CAS} is selected to keep n_2 in saturation. Small signals are shown for i_0, v_{in}, and v_{out}. The signal flow graph for this circuit is shown in Fig. 6.4b with input signals i_0 and v_{in} so that we can solve for v_{out} for either of these input signals. If we set v_{in} to zero and solve for v_{out}/i_0, we get the port impedance z_{out}, Eq. (6.4). The output impedance is increased by a gain factor $r_{ds2}g_m$ over the single transistor impedance and is a primary reason for using the cascode configuration.

It is interesting to note that the feedback here is positive. The loop is internal to the transistor so that we do not have access to measuring it on the bench or in simulation using the z-method except for hand analysis using the zero-order model as used here. The loop gain function is always less than one, making the configuration stable for positive feedback.

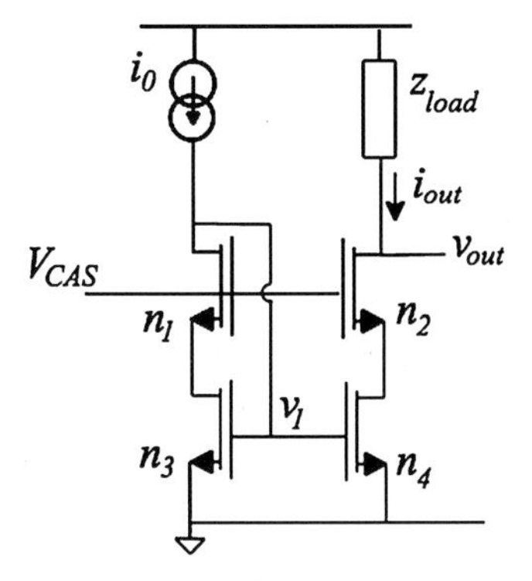

Fig. 6.3 Cascoded current mirror to remove output voltage variation in mirrored current

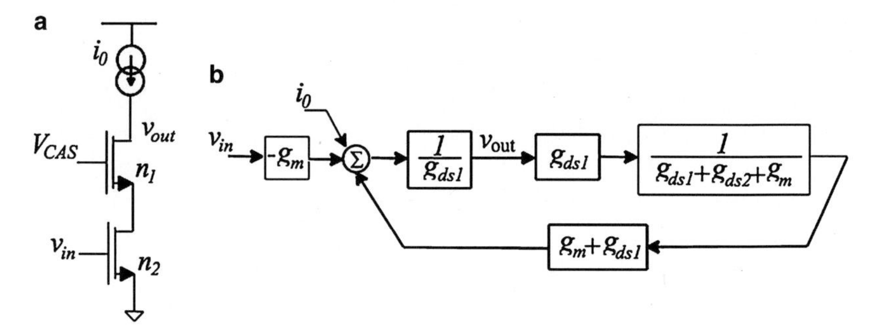

Fig. 6.4 (**a**) Cascode configuration for increasing output impedance and (**b**) flow graph for cascode output impedance

$$\begin{aligned} z_{\mathrm{out}} &= \frac{v_{\mathrm{out}}}{i_0} = \frac{\dfrac{1}{g_{\mathrm{ds1}}}}{1 - \dfrac{1}{g_{\mathrm{ds1}} + g_{\mathrm{ds2}} + g_{\mathrm{m}}} \cdot (g_{\mathrm{m}} + g_{\mathrm{ds1}})} \\ &= \frac{\dfrac{1}{g_{\mathrm{ds1}}} \cdot (g_{\mathrm{ds1}} + g_{\mathrm{ds2}} + g_{\mathrm{m}})}{g_{\mathrm{ds2}}} \\ &= r_{\mathrm{ds1}} + r_{\mathrm{ds2}} + r_{\mathrm{ds2}} . r_{\mathrm{ds2}} \cdot g_{\mathrm{m}} \end{aligned} \tag{6.4}$$

The flow graph also yields the output/input voltage transfer function. We have already solved for z_{out}. The small signal short-circuit current is $-g_m \times v_{in}$ and the output transfer function is $-g_m \times z_{out}$, Eq. (6.4).

$$H_{\mathrm{out}}(s) = \frac{v_{\mathrm{out}}}{v_{\mathrm{in}}} = (-g_{\mathrm{m}})(r_{\mathrm{ds1}} + r_{\mathrm{ds2}} + r_{\mathrm{ds2}} r_{\mathrm{ds2}} g_{\mathrm{m}}) \tag{6.5}$$

The Enhanced Cascode

An enhancement of this circuit is shown in Fig. 6.5a. The voltage v_{g1} (here allowed to float—we do not apply an auxiliary source to this node for this graph) is driven to $-A \times v_s$ by the feedback loop, making $v_{gs} = -v_s \times (A+1)$ and effectively multiplying g_m by $(A+1)$ in the results above for $z_0{}'$ and $H_{os}(s)$. The short-circuit current at node v_{out} due to v_{in} is added by inspection to the graph, arguing that the current into the drain of n_2 fully flows into the drain of n_1. By floating the gate of n_1, we in essence collapse two feedback loops due to the amplifier and n_1 in our mind simplifying the flow graph. This is not needed of course and can be done with an expanded graph showing the three loops in this feedback circuit, Fig. 6.6.

These graphs are not intended to yield the loop gain function. They couple partial circuit transfer relations algebraically only. We can now perform graph

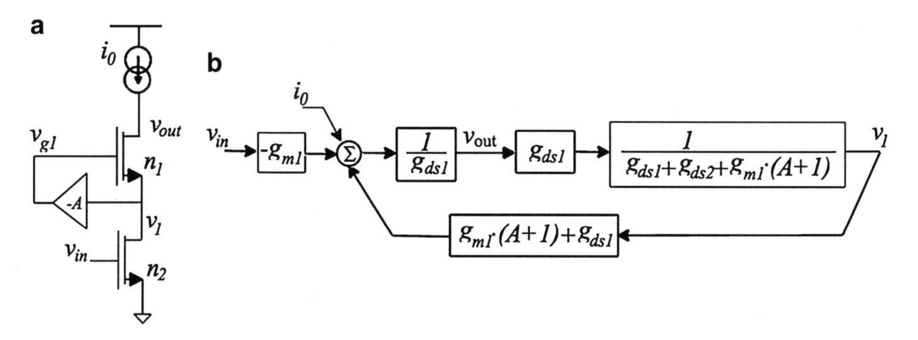

Fig. 6.5 (**a**) Enhanced g_m cascode-amplifying stage and (**b**) signal flow graph for port impedance at v_{out} and the voltage transfer function

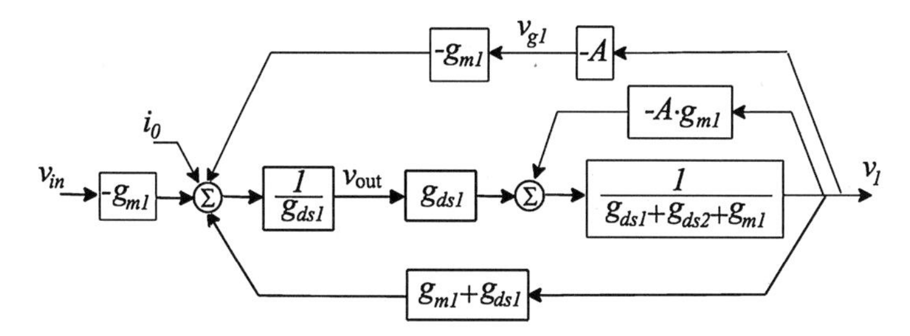

Fig. 6.6 Flow graph using the 2 internal nodes, v_1 and v_{g1}, showing 3 feedback loops

reduction manipulations, maintaining terminal relations. First we collapse the inner loop having the $(-A \times g_m)$ factor to obtain relation (Eq. 6.5):

$$\frac{1}{g_{\text{ds1}} + g_{\text{ds2}} + g_{\text{m}} \cdot (A+1)} \tag{6.6}$$

The graph now has a center forward branch with no loops and upper and lower feedback branches that are fully on the same level—they have the same input signal v_1 and add to the same summer so they can be algebraically combined into one feedback path fully reproducing that in Fig. 6.5 making the three-loop graph into one with a single loop. The amplifier is seen to effectively enhance the transconductance, g_m, by the factor $(A+1)$.

The Wilson Current Mirror

The Wilson current mirror, Fig. 6.7a, has this enhanced g_m design. The amplifier is composed of the transistor n_3 and its output impedance $1/g_{ds3}$, making $A = g_{m3}/g_{ds3}$. Transistor n_2 here is diode connected making its net impedance looking into the

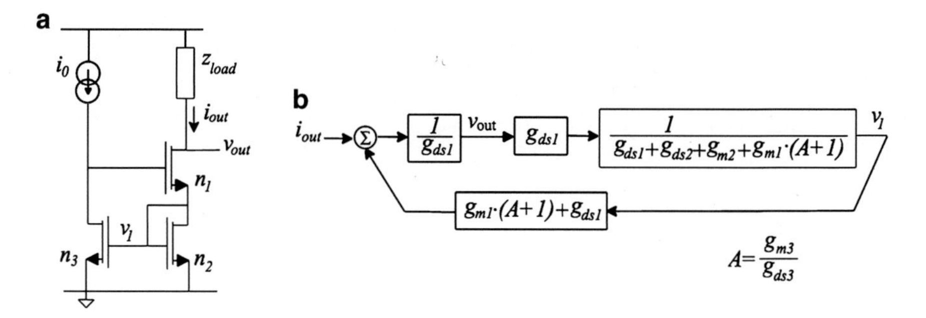

Fig. 6.7 (**a**) Schematic for the Wilson current mirror and (**b**) signal flow graph for the output impedance of the Wilson current mirror

drain/gate port $1/(g_{ds2}+g_{m2})$, an impedance lower than r_{ds2} in Fig. 6.5a. The output impedance for this cell is derived as the last line in Eq. (6.7) and is seen to be the same as that in Eq. (6.4) if we substitute the diode-connected nMOS impedance for r_{ds2}.

$$
\begin{aligned}
z_{\text{out}} = \frac{v_{\text{out}}}{i_{\text{out}}} &= \frac{\dfrac{1}{g_{\text{ds1}}}}{1-\dfrac{g_{\text{m1}}\cdot\left(1+\dfrac{g_{\text{m3}}}{g_{\text{ds3}}}\right)+g_{\text{ds1}}}{g_{\text{ds1}}+(g_{\text{ds2}}+g_{\text{m2}})+g_{\text{m1}}\cdot\left(1+\dfrac{g_{\text{m3}}}{g_{\text{ds3}}}\right)}} \\
&= \frac{\dfrac{1}{g_{\text{ds1}}}\left(g_{\text{ds1}}+(g_{\text{ds2}}+g_{\text{m2}})+g_{\text{m1}}\cdot\left(1+\dfrac{g_{\text{m3}}}{g_{\text{ds3}}}\right)\right)}{g_{\text{ds1}}+(g_{\text{ds2}}+g_{\text{m2}})+g_{\text{m1}}\cdot\left(1+\dfrac{g_{\text{m3}}}{g_{\text{ds3}}}\right)-\left(g_{\text{m1}}\cdot\left(1+\dfrac{g_{\text{m3}}}{g_{\text{ds3}}}\right)-g_{\text{ds1}}\right)} \\
&= \frac{\dfrac{1}{g_{\text{ds1}}}\left(g_{\text{ds1}}+g_{\text{ds2}}+g_{\text{m2}}+g_{\text{m1}}\cdot\left(1+\dfrac{g_{\text{m3}}}{g_{\text{ds3}}}\right)\right)}{g_{ds2}+g_{m2}} \\
&= r_{\text{ds1}}+\frac{1}{g_{\text{ds2}}+g_{\text{m2}}}+\frac{r_{\text{ds1}}}{g_{\text{ds2}}+g_{\text{m2}}}g_{\text{m1}}\cdot\left(1+\frac{g_{\text{m3}}}{g_{\text{ds3}}}\right)
\end{aligned}
\tag{6.7}
$$

Returning to the Wilson current mirror schematic in Fig. 6.7 we see that the current transfer function i_{out}/i_0 is the negative of the short-circuit current at the output node multiplied by the current divider between the load and the port impedance looking into the drain of n_1, Eq. (6.8). While we can build the flow graph for this variable, a more direct and simpler graph can be used if we recognize that current i_{out} is the same as the current into diode n_2, a function of v_1 and the transistor n_2. The flow graph for v_1 is shown in Fig. 6.8 and, assuming y_{load} is greater than y_{out}, $i_{out}=-i_{sc,out}$.

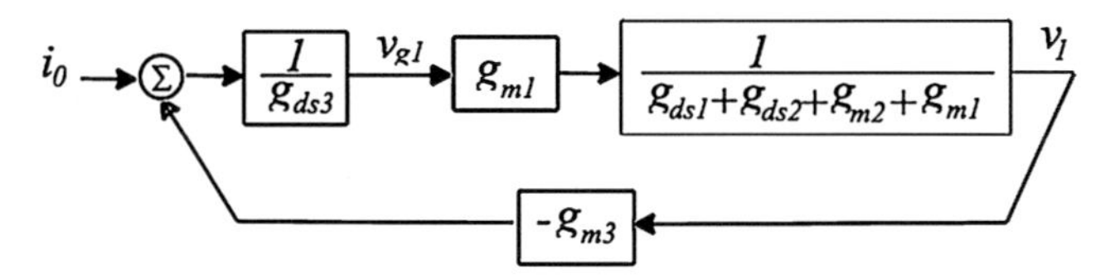

Fig. 6.8 v_1/i_0 transfer function flow graph for the Wilson current mirror holding v_1 at ac ground

$$\begin{aligned} i_{\text{out}} &= -i_{\text{sc,out}} \cdot \frac{y_{\text{load}}}{y_{\text{out}} + y_{\text{load}}} \\ \sim -i_{\text{sc,out}} &= \left(g_{\text{m2}} + g_{\text{ds2}}\right) \cdot v_1 \end{aligned} \tag{6.8}$$

Transistors n_2 and n_3 are by design matched in layout and current, making their model parameters equal. Solving for v_1 and using Eq. (6.8), we get i_{out}, Eq. (6.8), equal to the input current i_0.

$$\begin{aligned} i_{\text{out}} &= \frac{r_{\text{ds3}} \cdot g_{\text{m1}} \dfrac{1}{g_{\text{ds1}} + g_{\text{m1}} + g_{\text{ds2}} + g_{\text{m2}}}}{1 + r_{\text{ds3}} g_{\text{m3}} g_{\text{m1}} \dfrac{1}{g_{\text{ds1}} + g_{\text{m1}} + g_{\text{ds2}} + g_{\text{m2}}}} \left(g_{\text{ds2}} + g_{\text{m2}}\right) i_0 \\ &\sim i_0 \end{aligned} \tag{6.9}$$

The Single-Stage Differential Operational Transconductance Amplifier (OTA)

A primary building block in CMOS amplifier design is the single-stage operational transconductance amplifier, OTA, using a current mirror active load and a current mirror cell bias, Fig. 6.9. DC reference current I_0 is mirrored through transistor n_1 to make the drain current of n_2 approximately equal to I_0. This current controls the operating conditions of the OTA.

The difference voltage applied to the cell, $v_p - v_n$, drives a current onto node v_{out} to produce the output voltage response. We start with the zero-order analysis using simplifying approximations along with the DPI/SFG tool. Using the differential mode input signal, we apply $+v_d$ to v_p and $-v_d$ to v_m, where $v_d = (v_p - v_m)/2$. We then approximate the response of the p-load current mirror to be ideal, $H_{curr\_mirr} = 1$, that the drain current in transistor n_3 is mirrored onto the drain node of n_4. The flow graph for the short-circuit current at the output is shown in Fig. 6.10.

Assuming unity for the pMOS load current mirror, we write for $i_{sc,out}$, Eq. (6.10):

$$\begin{aligned} i_{\text{sc,out}} &= \left(1 + H_{\text{curr\_mirr}}\right) \cdot g_{\text{mn}} \cdot v_{\text{d}} \\ &= g_{\text{mn}} \left(v_{\text{p}} - v_{\text{m}}\right) \end{aligned} \tag{6.10}$$

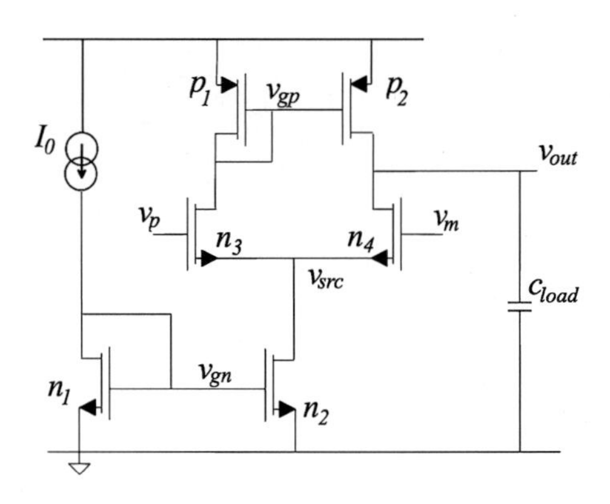

Fig. 6.9 Single-stage OTA using current mirror active load

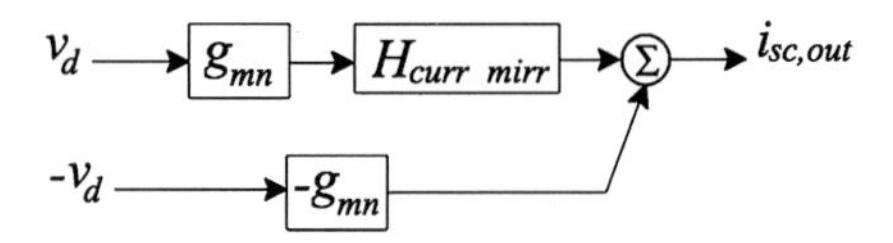

Fig. 6.10 Signal flow graph for output short-circuit current for the single-stage OTA

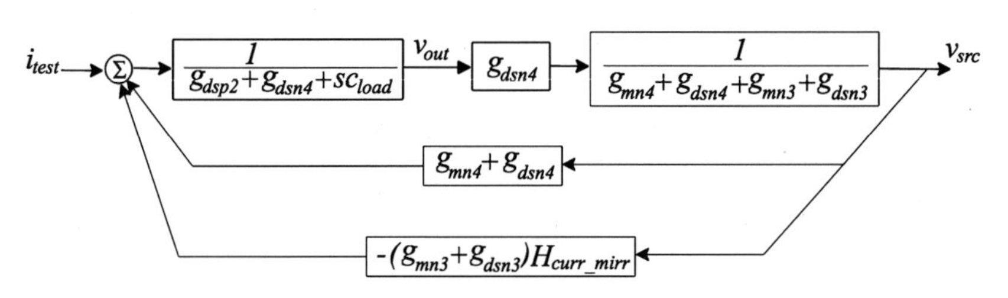

Fig. 6.11 Signal flow graph for output DPI for the single-stage OTA

The output DPI flow graph is shown in Fig. 6.11. Here we see a part of the flow graph reflecting that of the cascode impedance graph, the positive loop gain portion. A negative feedback path through the pMOS current mirror cancels this positive feedback term to the extent that the current mirror load transfer function is in unity. The effective output impedance is approximately the parallel of the output impedances of m_{n4}, m_{p2}, and the load capacitance, c_{load}, a single-pole response.

$$\begin{aligned} \mathrm{dpi}_{\mathrm{out}} &= \frac{1}{g_{\mathrm{dsp2}} + g_{\mathrm{dsn4}} + sc_{\mathrm{load}}} \\ H_{\mathrm{out}} &= g_{\mathrm{mn}} \frac{1}{g_{\mathrm{dsp2}} + g_{\mathrm{dsn4}} + sc_{\mathrm{load}}} \end{aligned} \tag{6.11}$$

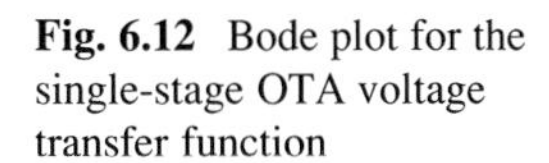

Fig. 6.12 Bode plot for the single-stage OTA voltage transfer function

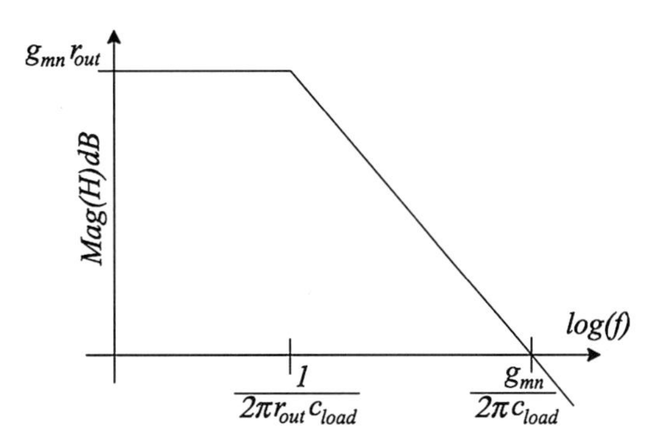

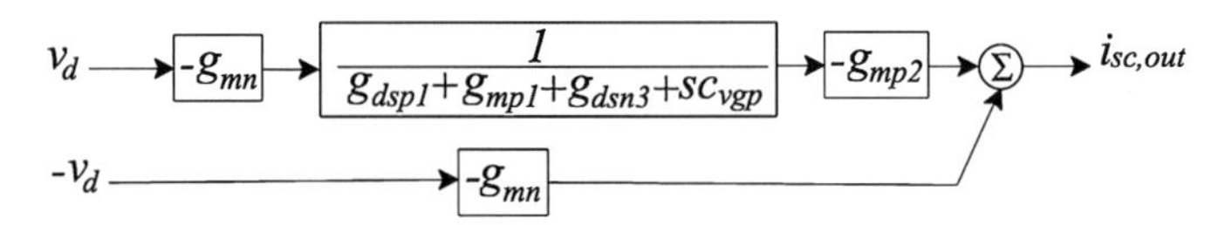

Fig. 6.13 Flow graph for the OTA short-circuit current including the nodal impedance at v_{gp}

Writing the real part of the output impedance as $r_{out} = 1/(g_{dsp2} + g_{dsn4})$, we can create the Bode plot for this OTA as shown in Fig. 6.12. The low-frequency gain is found to be $g_{mn} \times r_{out}$, and the gain-bandwidth product is g_{mn}/c_{load} in radians per second.

Let us examine the assumed unity transfer function for the current mirror load of the OTA. Node v_{gp} is a relatively low impedance node due to the diode-connected pMOS m_{p1}. This node will also have a capacitance associated with it due to device parasitics adding a pole to the response at this node. Adding these elements to the analysis will primarily affect the output short-circuit current. The output DPI will remain essentially as Eq. (6.11). The assumed unity current mirror transfer function H_{curr_mirr} becomes as shown in Fig. 6.13. The capacitance at this node consists of the gate capacitances of m_{p1} and m_{p2} plus drain diffusion capacitances of m_{p1} and m_{n3}. The nodal resistance is dominated by g_{mp1}, making $i_{sc,out}$ as given in Eq. (6.12) after dropping the g_{ds} terms and equating g_{mp1} to g_{mp2}. We see that the response at this node adds a pole at frequency $f_{pvgp} = g_{mp1}/(2\pi\ c_{vpg})$ and a zero at twice this frequency. The pole is at a relatively high frequency and can be designed to be above the unity gain frequency of the loaded OTA. Its effects are moderated by zero so that its impact on the overall transfer function is modest—a reduction in gain of a factor of 2 and a phase dip and recovery around these frequencies.

$$
\begin{aligned}
i_{\mathrm{sc,out}} &= g_{\mathrm{mn}}\left(1+\frac{1}{g_{\mathrm{dsp1}}+g_{\mathrm{dsn3}}+g_{\mathrm{mp1}}+sc_{\mathrm{vgp}}}\cdot g_{\mathrm{mp2}}\right)\\
&\cong g_{\mathrm{mn}}\left(1+\frac{\frac{1}{g_{\mathrm{mp1}}}}{1+\frac{sc_{\mathrm{vgp}}}{g_{\mathrm{mp1}}}}\cdot g_{\mathrm{mp2}}\right)\\
&= g_{\mathrm{mn}}\left(\frac{2+\frac{sc_{\mathrm{vgp}}}{g_{\mathrm{mp1}}}}{1+\frac{sc_{\mathrm{vgp}}}{g_{\mathrm{mp1}}}}\right)
\end{aligned}
\tag{6.12}
$$

The output/input response becomes as shown in Eq. (6.13).

$$
\begin{aligned}
H_{\mathrm{out}}(s) &= g_{\mathrm{mn}}\left(\frac{2+\frac{sc_{\mathrm{vgp}}}{g_{\mathrm{mp1}}}}{1+\frac{sc_{\mathrm{vgp}}}{g_{\mathrm{mp1}}}}\right)\frac{1}{g_{\mathrm{dsp2}}+g_{\mathrm{dsn4}}+sc_{\mathrm{load}}}\\
&\cong g_{\mathrm{mn}}r_{\mathrm{out}}\left(\frac{2+\frac{sc_{\mathrm{vgp}}}{g_{\mathrm{mp1}}}}{1+\frac{sc_{\mathrm{vgp}}}{g_{\mathrm{mp1}}}}\right)\frac{1}{1+sc_{\mathrm{load}}r_{\mathrm{out}}}\\
r_{\mathrm{out}} &= \frac{1}{g_{\mathrm{dsp2}}+g_{\mathrm{dsn4}}}
\end{aligned}
\tag{6.13}
$$

Figure 6.14 shows the ac response for the OTA (0.18 μm technology with *10 pF* load biased at *2 μA* with $g_{mn}=24.2\ \mu S$). Signal v_{out} shows a low-frequency gain of *50 dB*, a pole at about *1 kHz*, and a $GBW=385\ kHz$. The pole at the current mirror load, v_{gp}, is at about *2.3 MHz*. i_{sc}, the signal at the bottom of the plot, shows a *−6 dB* step due to this pole and the zero at twice this frequency. This pole-zero pair adds some structure to the response of the OTA but usually is not an issue.

Reference Current and Voltage Generators

Electronic systems require references, currents and voltages, in order to establish an operating point or a comparison standard. A constant voltage is used in analog-to-digital voltage conversions, for example, to provide a scale against which a sampled signal is compared for digital representation. Reference currents are mirrored and scaled to provide operating bias levels of amplifiers to keep the design within performance and power specifications. A simple current generator is shown in Fig. 6.15, usable where the current I_{out} is not required to be very accurate.

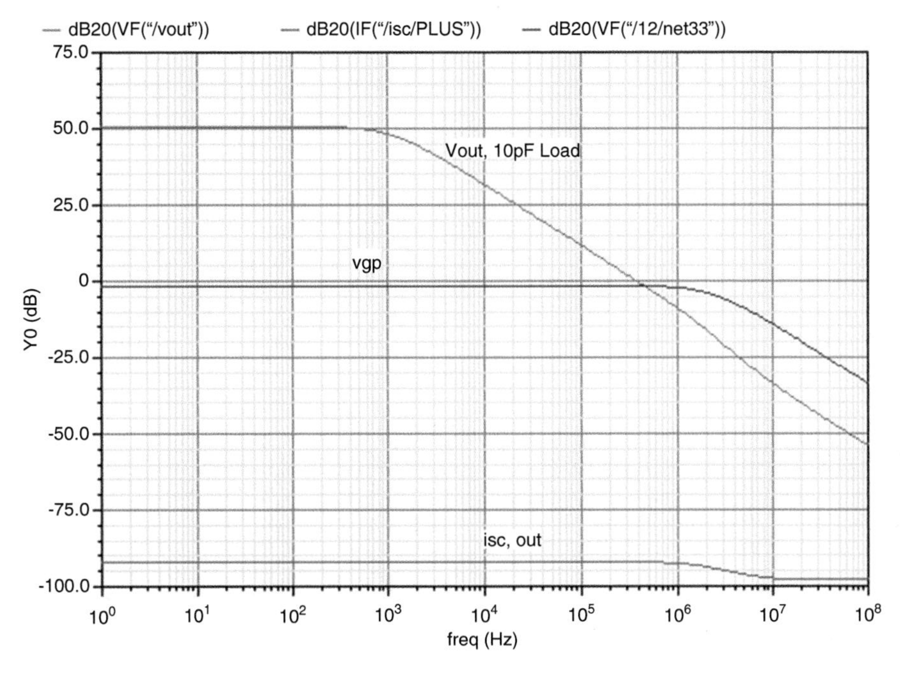

Fig. 6.14 AC simulation responses for single-stage OTA—v_{out}, v_{gp}, and i_{sc}

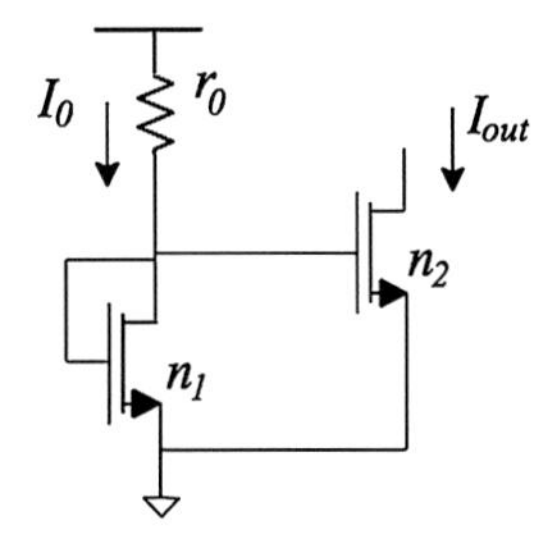

Fig. 6.15 Simple current generator

Constant Current Generator Using an OTA and Feedback

In Fig. 6.12, we see that the OTA's low-frequency gain and its GBW are direct functions of transistor parameters g_m and g_{ds}, which are direct functions of bias current. To control the OTA performance, we need to provide it with a better controlled reference current than that in Fig. 6.15. The circuit in Fig. 6.16 uses a feedback loop to control the voltage on resistor r_{set} equal to a supplied reference voltage V_{ref}. The current through this resistor is then passed through a current mirror where it can source this current to a load.

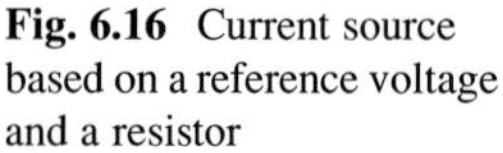

Fig. 6.16 Current source based on a reference voltage and a resistor

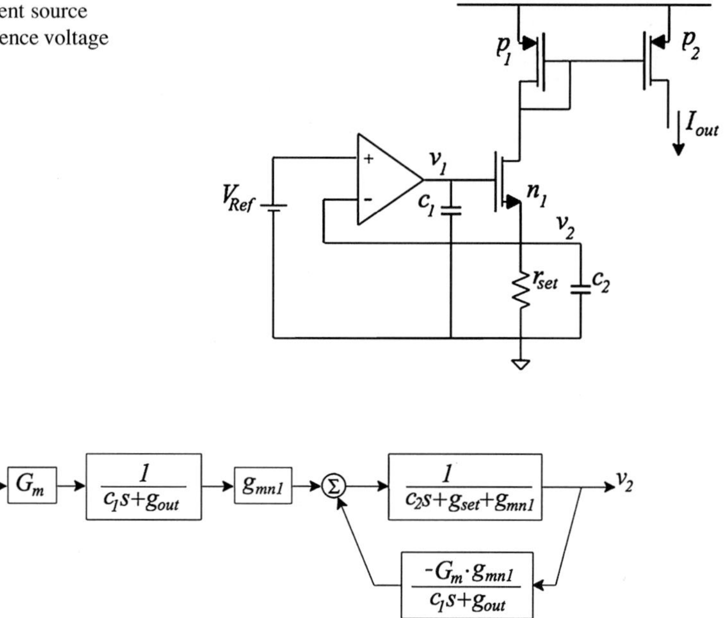

Fig. 6.17 Flow graph for current source in Fig. 6.16

Using the OTA from Fig. 6.9 (and the idealized load mirror approximation), we derive the flow graph for this loop as shown in Fig. 6.17. Using this graph, we solve for the loop gain and for node voltage v_2 to obtain the output current, i_{out}, Eq. (6.14), where G_m is the OTA transconductance, g_{out} is the output conductance of the OTA, and c_1 and c_2 are the net capacitance at nodes v_1 and v_2.

$$
\begin{aligned}
\text{loop gain} &= G_{\mathrm{m}} \frac{1}{c_1 s + g_{\mathrm{out}}} g_{\mathrm{mn1}} \frac{1}{c_2 s + g_{\mathrm{set}} + g_{\mathrm{mn1}}} \\
v_2(s) &= \frac{G_m \dfrac{1}{c_1 s + g_{out}} g_{mn1} \dfrac{1}{c_2 s + g_{\mathrm{set}} + g_{\mathrm{mn1}}}}{1 + G_{\mathrm{m}} \dfrac{1}{c_1 s + g_{\mathrm{out}}} g_{\mathrm{mn1}} \dfrac{1}{c_2 s + g_{\mathrm{set}} + g_{\mathrm{mn1}}}} \\
i_{\mathrm{out}}(s) &= v_2 g_{\mathrm{set}} = V_{\mathrm{ref}} g_{\mathrm{set}}
\end{aligned}
\tag{6.14}
$$

We see that the loop gain has two poles, one at a high impedance node at the output of the OTA and the other at the source of transistor n_2, a low impedance node. Compensation can easily be achieved by adding capacitance to node v_1. The generated current I_{out} can be multiplied by paralleling output transistors. The generated voltage at the gate of transistor p_1 can drive multiple mirroring devices to provide additional reference currents. The resulting currents are dependent on V_{ref} and on r_{set}. These can be supplied externally as precision off-chip resistors and reference voltages, or they can be generated on chip and individually trimmed to meet a specification. Note that if an on-chip r_{set} is used, the generated current will reflect the particular chip-processing variation so that if this current is mirrored onto another resistor of the same type, the temperature and processing variation will track. A voltage generated using another resistor will have these variations cancel to first order.

The Constant G_m Current Generator: A Positive Feedback Loop

A popular current source is the "constant g_m" design shown in Fig. 6.18. In a double current mirror design using source degeneration and positive feedback, the currents in the core branches are forced equal. This current is then mirrored out in either p-type or n-type mirrors. Nodes v_1 and v_2 are available for mirroring using nMOS (pMOS) transistors matching device n_1 (p_1). Caution: Circuits such as this one that do not have inputs may fail to start—the circuit has two stable operating points, the one we want plus a non-energized state with the branch currents both zero. For such a circuit, we need to add a start-up circuit, such as that shown in the dashed box, that provides a transient pulse until the feedback takes over. Start-up circuits generally require a current that is always on and an injection point that can be turned off once the primary circuit becomes active. Some capacitive-based start-up circuits exist that provide a single transient start cycle and do not draw steady-state current. These need to be used with caution as you only get the one shot.

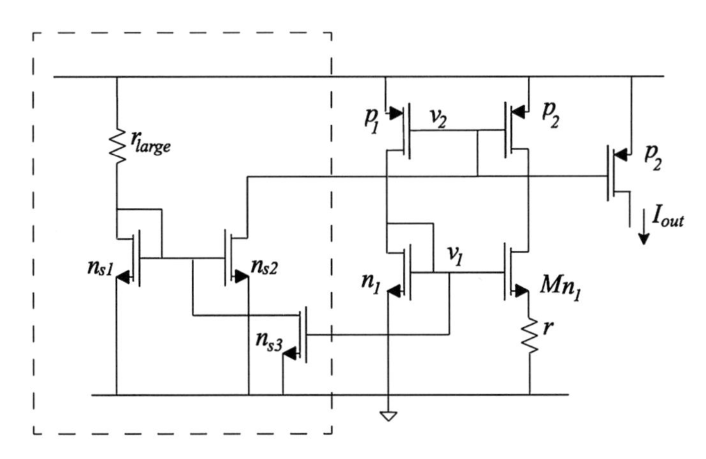

Fig. 6.18 "Constant g_m" current source

Large-signal analysis using the square low model ($I_{ds}=K(V_{gs}-V_{th})^2$) for the transistors reveals that the operating transconductance of the nMOS transistors will be proportional to the $1/r=g$, the conductance at the source of M_{n1}, hence the name of the cell. The voltage across the resistor, v_r, is found as in Eq. (6.15):

$$\begin{aligned} V_{\mathrm{r}} &= V_{\mathrm{gsn1}} - V_{\mathrm{gsMn1}} \\ &= \sqrt{\frac{I_{\mathrm{ds}}}{K_{\mathrm{n1}}}}\left(1-\frac{1}{\sqrt{M}}\right) \end{aligned} \tag{6.15}$$

Equating the resistor current to the current through n_1, we get Eq. (6.16) where the back gate effect in device M_{n1} is ignored

$$I_{\mathrm{r}} = \sqrt{\frac{I_{\mathrm{ds}}}{K_{\mathrm{n1}}}}\left(1-\frac{1}{\sqrt{M}}\right)g = K_{\mathrm{n1}}\left(V_{\mathrm{gsn1}} - V_{\mathrm{thn1}}\right)^2 \tag{6.16}$$

Substituting for $g_m=2K(V_{gs}-V_{th})$ in this relation results in Eq. (6.17). It is popular to use $M=4$ in this relation, making the transistor's transconductance $g_{mn1}=g$. Mirroring this current onto other nMOS devices will set the transconductance of those devices as well.

$$g_{\mathrm{mn1}} = \left(1-\frac{1}{\sqrt{M}}\right)\frac{g}{2} \tag{6.17}$$

The transconductance of the pMOS devices will also be proportional to g as shown in Eq. (6.18).

$$g_{\mathrm{mp}} = \frac{\sqrt{K_{\mathrm{p}}}}{\sqrt{K_{\mathrm{n1}}}}\left(1-\frac{1}{\sqrt{M}}\right)\frac{g}{2} \tag{6.18}$$

The loop gain for this system is interesting in that the loop has positive feedback. This can be seen in the flow graph in Fig. 6.19. A test current is injected into the v_1 node and the loop generated following the z-method for loop gain analysis developed in Chap. 5. The crosscurrent here assumes ideal pMOS current mirroring and uses the source-degenerated transconductance for n_2, and the flow graph is drawn by inspection of the circuit. The loop gain is found to be as given in Eq. (6.19) where we dropped g_{ds} terms compared to g_m ones.

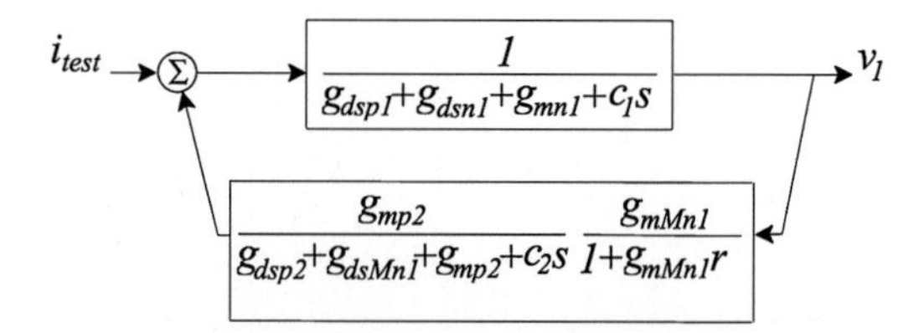

Fig. 6.19 Loop gain flow graph for "constant g_m" current source

$$\begin{aligned}\text{loop gain} &= \frac{g_{\text{mMn1}}}{1+g_{\text{mMn1}}r}\,\frac{g_{\text{mp2}}}{g_{\text{mp2}}+c_2 s}\,\frac{1}{g_{\text{mn1}}+c_1 s}\\ &\cong \frac{g_{\text{mMn1}}}{1+g_{\text{mMn1}}r}\,\frac{1}{g_{\text{mn1}}}\end{aligned} \tag{6.19}$$

Using 0.18 μm technology transistors with $L = 0.2\ um$, $W = 4\ um$ and $M = 4$, and $r = 100\ K$, we get a loop gain of ~-9.36 dB (0.34), top plot in Fig. 6.20, having approximately 0° phase up to ~*10 MHz* as shown in the bottom plot of this figure.

With a loop gain of ~*0.34*, we have a stable system, but we also have a relatively large gain for mismatch effects. Injecting a mismatch current at either of the nodes v_1 or v_2 will result in a change in the core current of *50 %* of the mismatch current, Eq. (6.1):

$$\begin{aligned}\text{Delta} - I &= \frac{I_{\text{mismatch}}}{1 - \text{loopgain}} = \frac{I_{\text{mismatch}}}{1 - 0.34}\\ &= I_{\text{mismatch}} \cdot 1.51\end{aligned} \tag{6.20}$$

Simulation results with *0, 100*, and *−100 nA* injected offset currents resulted in the following pMOS and nMOS unit currents (Table 6.1):

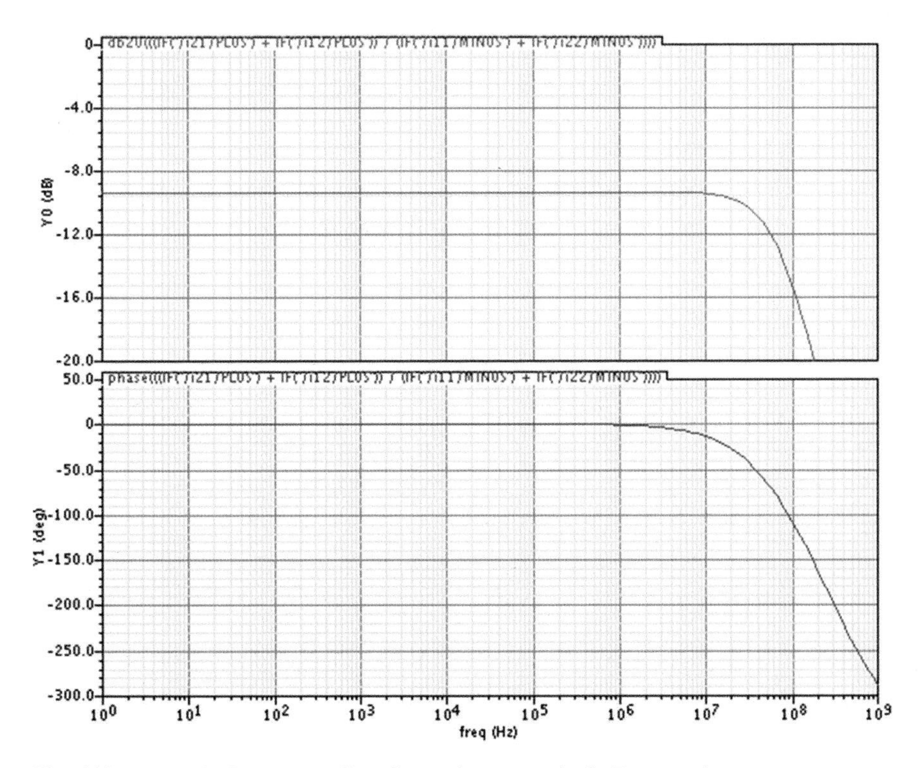

Fig. 6.20 Loop gain for constant G_m cell, top shows magnitude, Bottom, phase

Table 6.1 Calculated results for offset currents in constant G_m cell using Eq. (6.20)

i_{offset}	i_p	i_n
0	$603n$	$713n$
$100n$	$442n$	$527n$
$-100n$	$747n$	$878n$

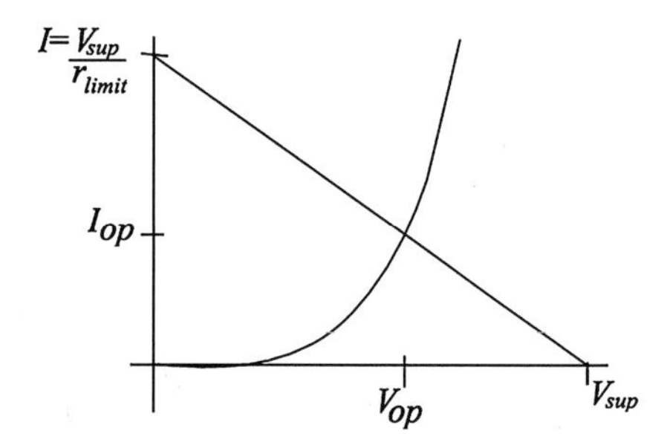

Fig. 6.21 Load line plot for operating point of series resistor plus nonlinear device

A *5 mV* threshold mismatch, for example, in a design using small geometries with $g_m = 16\,\mu$S, the mismatch current is *80 nA* for 1-sigma. This variation can be reduced by proper transistor sizing as discussed in Chap. 4 on matching.

Voltage References

Voltage references, as we saw with current references, can be simple and provide a reference that is not very accurate and may have large temperature and manufacturing dependencies. Such designs depend on nonlinear device behavior such as a breakdown voltage (Zener diode), a forward-biased junction, or a turn-on voltage (threshold voltage) of an active device. The operating point for circuits having a resistor and a nonlinear device in series across a power source can be found using a load-line graph as in Fig. 6.21. The voltage across the devices in series sum to the supply voltage and the current through each is the same. The intersection of the linear resistor line with that of the nonlinear device defines the operating point. A few such circuits are shown in Fig. 6.22.

Elements of the Bandgap Reference Voltage

A reference voltage that is stable with temperature and process variations is needed in many analog systems. The bandgap reference voltage generator meets this requirement by summing a voltage that varies negatively with temperature with

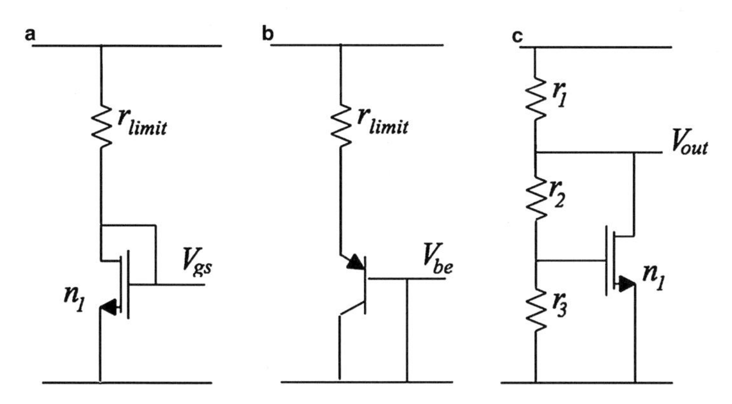

Fig. 6.22 Simple reference voltage generators, (**a**) V_{gs} voltage, (**b**) diode voltage, and (**c**) V_{gs} multiplier

one that varies positively. The forward voltage of a diode at a given forward current I_f has a negative temperature coefficient. This forward voltage can be written as in Eq. (6.21) where the "1" is ignored compared to the exponential with positive V_{be}.

$$I_{\mathrm{f}} = I_{\mathrm{s}}\left(e^{\frac{q}{kT}V_{\mathrm{be}}} - 1\right)$$
$$V_{\mathrm{be}} = \frac{kT}{q}\ln\left(\frac{I_{\mathrm{f}}}{I_{\mathrm{s}}}\right) \tag{6.21}$$

The diode voltage temperature dependence is dominated by the temperature response of the saturation current I_s so that the net dependence is negative and of the order of $-2\ mV/°K$ [1, 2].

Taking two diodes with different current densities, usually by using either the same current in diodes having different areas or diodes with the same area but with different currents and subtracting their forward voltages, we remove the saturation current effect and obtain a signal with *positive* temperature coefficient:

$$\Delta V_{\mathrm{be}} = V_{\mathrm{be2}} - V_{\mathrm{be1}} = \frac{kT}{q}\ln\left(\frac{I_{\mathrm{f2}}}{I_{\mathrm{s2}}}\ \frac{I_{\mathrm{s1}}}{I_{\mathrm{f1}}}\right)$$
$$= \frac{kT}{q}\ln\left(\frac{I_{\mathrm{f2}}}{I_{\mathrm{f1}}}\ \frac{A_1}{A_2}\right). \tag{6.22}$$

If we use the same current through two diodes with area ratio M, the larger device will develop a smaller forward voltage. Imposing this voltage difference on a resistor will produce a current that is PTAT, proportional to absolute temperature, T, in degrees Kelvin. Consider the circuit shown in Fig. 6.23 consisting of two branches biased with identical currents I_{cross}. One branch has a unit diode, while the other has M unit diodes in parallel plus a resister in series with the current source.

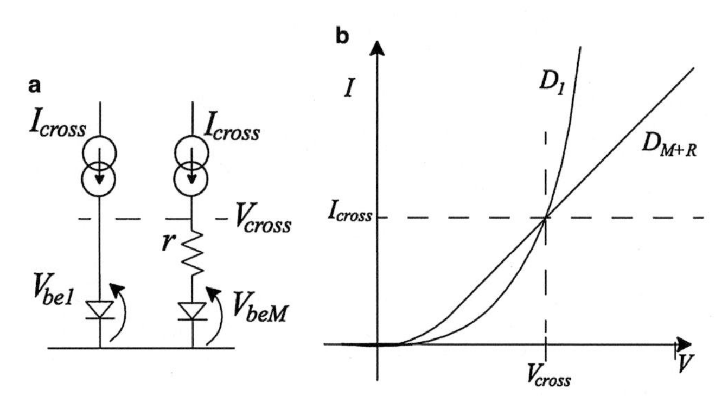

Fig 6.23 (**a**) ΔV_{be} generator and (**b**) operating point for where both branches have current I_{cross} and voltage V_{cross}

The *I–V* sketch in Fig. 6.23b shows the branch currents as the ordinate and the voltages as the abscissa for the single diode and for the *M*-parallel plus series resistor combination. At low current, $V_{beM}+I\times r$ is less than $V_{\mathrm{be}1}$. As the current is increased, the voltage across the resistor increases so that the forward bias on the *M*-parallel diodes becomes less than that of the single diode with the same current. This continues with increasing current until the plots cross at the point the voltages are equal, $V_{beM}+I\times r=V_{be1}$. This intersection condition occurs at I_{cross} defined as in Eq. (6.23) where *T* is the temperature in degrees Kelvin, *k* is the Boltzmann constant (1.3806484×10^{-23} J/K), *q* (1.602×10^{-19} C) the electronic charge, and *M* is the number of parallel unit diodes in the large diode.

$$\mathrm{I_{cross}}=\frac{\Delta V_{\mathrm{be}}}{r}=\frac{kT}{q}\ln(M)\frac{1}{r} \tag{6.23}$$

With this cell, we have an element that will produce a decreasing voltage with increasing temperature, a diode, and can generate a voltage that will increase in magnitude with increasing temperature, ΔV_{be}. We need to create a situation where we combine these responses in the proper proportion so that one cancels the effect of the other. If we mix these two types of temperature response properly, we will get a relatively constant voltage over a useful range of temperature. This voltage is close to 1.2 V and is known as the bandgap voltage.

The feedback circuit in Fig. 6.24 replaces the current sources in Fig. 6.23a using two equal value resistors, r_2. The circuit will drive the OTA input voltages to the same value, making the voltages across the r_2 resistors equal to each other and therefore the diode branch currents equal to I_{cross}. Since this r_2 current is also the current through r_1, the current is defined as in Eq. (6.23) with $r=r_1$. The resulting voltage at the top of the r_2 resistors becomes as given in Eq. (6.24).

Fig. 6.24 Basic bandgap reference voltage generator

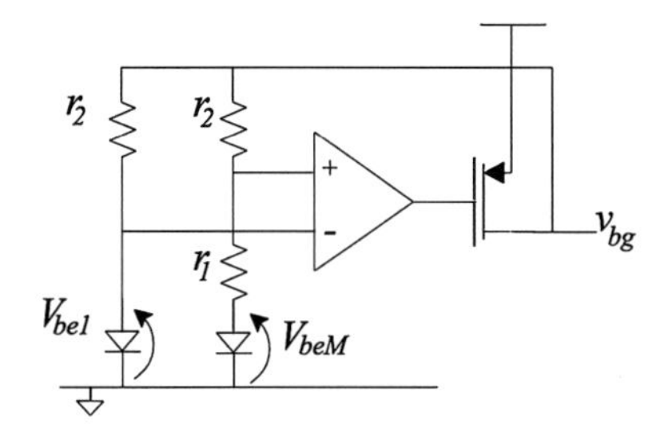

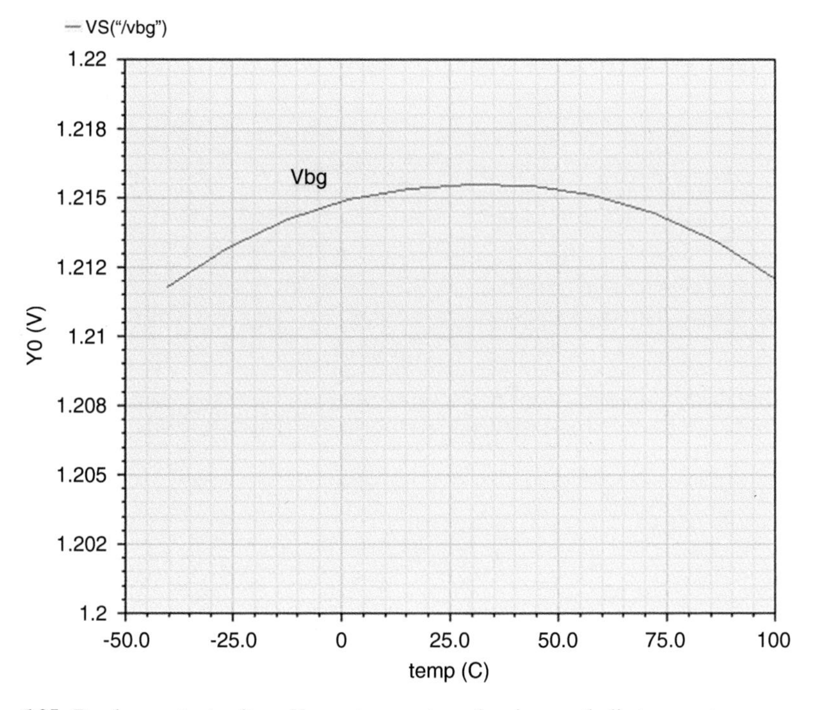

Fig. 6.25 Bandgap output voltage V_{bg} vs. temperature showing parabolic temperature response

$$\mathrm{V_{bg}} = V_{\mathrm{be1}}(I_{\mathrm{f}}, T) + \frac{kT}{q}\ln(M)\frac{1}{r_1}r_2 \tag{6.24}$$

Figure 6.25 shows simulated results for a bandgap reference voltage design having $M = 15$. V_{bg} is *1.215 V* at the peak—the bandgap voltage versus temperature shows parabolic curvature due to nonlinear behavior of the elements with temperature. The resistors are sized so that at 27 °C the diode branch current is about *220 nA* for low power operation. Here we see a ~*3 mV* variation in v_{bg} over a temperature change of about *140 °C*.

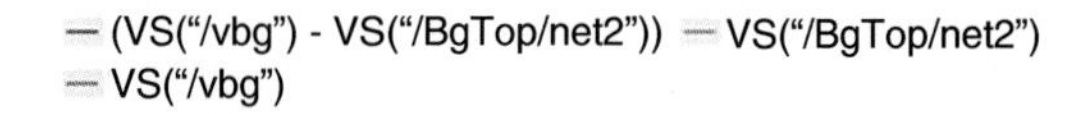

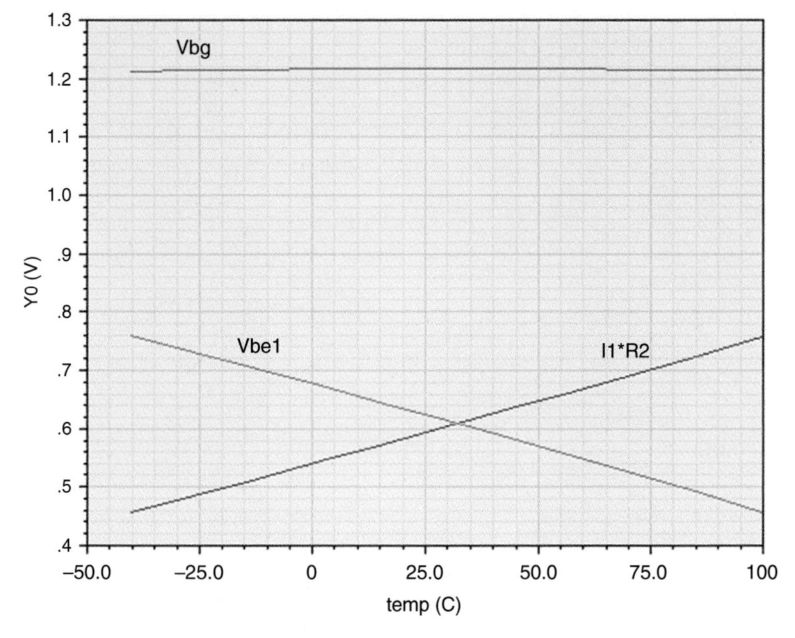

Fig. 6.26 Bandgap-simulated response showing V_{bg} (top) and the voltage components—V_{be1} with a negative temperature response and $I_1 \times r_2$ for the positive correction voltage

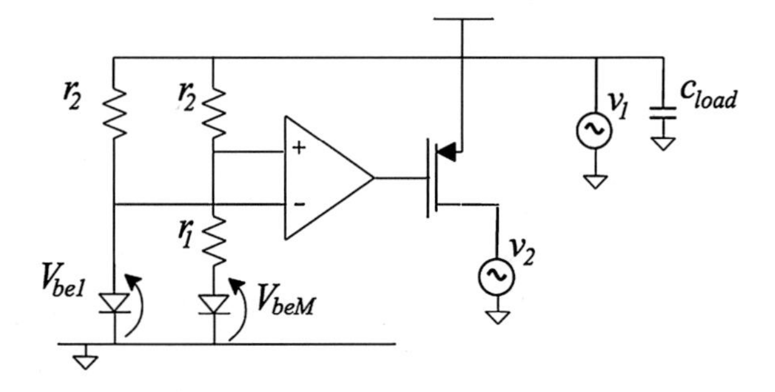

Fig. 6.27 Bandgap setup for loop gain analysis

Figure 6.26 shows the behavior of the components of the bandgap voltage V_{bg}, V_{be1} with a negative temperature coefficient and the voltage generated across one of the r_2 resistors due to the PTAT current I_1 through r_1. The sum of these voltages is V_{bg}, top trace in the plot.

The bandgap reference voltage system has two outer loops—one having negative feedback and the other positive. Taking the drain of the output pMOS transistor as our loop gain port, Fig. 6.27, we formulate the set of currents that define the loop gain function. The design is seen to have negligible reverse loop current, making $i_{21} \sim 0$. Assuming a load capacitance c_{load}, we write the remaining loop gain currents as shown in Eq. (6.25).

$$\text{Loop Gain} = \frac{-i_{21}}{i_{11}+i_{22}}$$

$$i_{21} = \left(\frac{r_{\rm d1}}{r_{\rm d1}+r_2} - \frac{r_{\rm dM}+r_1}{r_{\rm dM}+r_1+r_2}\right)(-G_{\rm m})\frac{1}{g_{01}+sc_{01}}g_{\rm mpout}v_{11}$$

$$i_{11}+i_{22} = \left(sc_{load} + \frac{1}{r_{\rm d1}+r_2} + \frac{1}{r_{\rm dM}+r_1+r_2}\right)v_{11} + g_{\rm dspout}v_{22} \quad (6.25)$$

$$\text{Loop Gain} = \frac{\left(\frac{r_{\rm d1}}{r_{\rm d1}+r_2} - \frac{r_{\rm dM}+r_1}{r_{\rm dM}+r_1+r_2}\right)G_{\rm m}\frac{1}{g_{01}+sc_{01}}g_{\rm mpout}}{\left(sc_{\rm load} + \frac{1}{r_{\rm d1}+r_2} + \frac{1}{r_{\rm dM}+r_1+r_2}\right)v_{11} + g_{\rm dspout}v_{22}}$$

The loop gain function, Eq. (6.25), is seen to have a minimum of two poles and therefore may become unstable. Here we see that we can add capacitance to the output node or to the OTA output node to gain phase margin.

This system, as with the constant g_m current generator, has two stable states and will require a start-up circuit to insure proper operation.

The Bandgap-Derived Constant Current Source

Another configuration of the elements of the core bandgap elements in Fig. 6.23a produces a constant current with low temperature dependence. Since the voltage across a unit diode can be made equal to that of a larger diode with a series resistance at a particular current, we can use mirrored output currents to drive each branch as shown in Fig. 6.28. Here the feedback maintains the OTA inputs at the voltage at the crossing of operating points of the single-unit diode voltage V_{be1} and that of the series r_1 and M-parallel diodes, V_{beM}. The current through the diodes

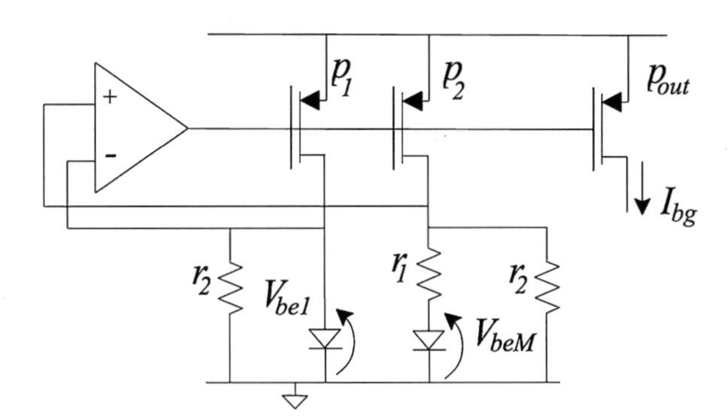

Fig. 6.28 Temperature-independent current generation using a bandgap configuration

is the PTAT current through r_1 having the difference in diode forward voltages across it, Eq. (6.26).

$$i_{r1} = \frac{kT}{q}\ln(M)\frac{1}{r_1} \tag{6.26}$$

The currents through transistors p_1 and p_2 will be as given in Eq. (6.27) where we have a sum of a PTAT current with one that is CTAT, complimentary to absolute temperature. By suitable selection of resistors r_1 and r_2 for a given technology, we can balance the temperature variations, making a current insensitive to temperature. This cell will also require a start-up circuit:

$$i_{p1} = i_{p2} = \frac{kT}{q}\ln(M)\frac{1}{r_1} + \frac{V_{be1}}{r_2} \tag{6.27}$$

Other Useful Cells

We have looked at some fundamental blocks that will appear in most designs—references, mirrors, and amplifiers. We now look at a sample of circuits and systems, analyzing basic operation using DPI/SFG techniques.

Capacitive Gain Stage

A gain stage with capacitive elements is useful in situations where ac coupling to the signal source is necessary such as in medical probing where we want very low power dissipation and band-pass operation. Very large resistors are used to set the DC conditions at the amplifier inputs and a very low lower cutoff frequency, Fig. 6.29.

Assuming that the amplifier is not the limiting element in the design and that it can be modeled as a transconductance G_m and output conductance g_{out}, we can map

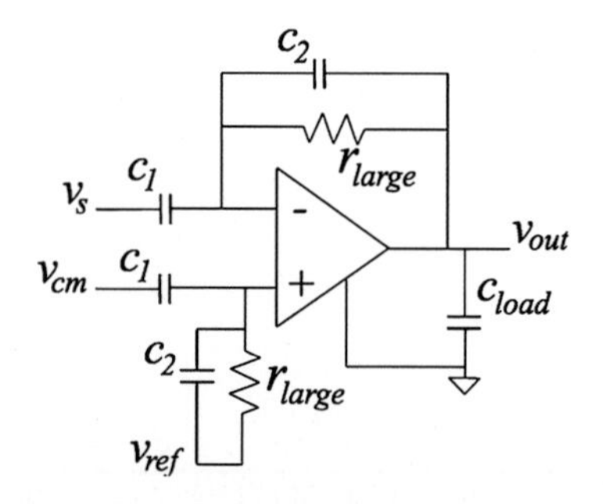

Fig. 6.29 Capacitive gain stage

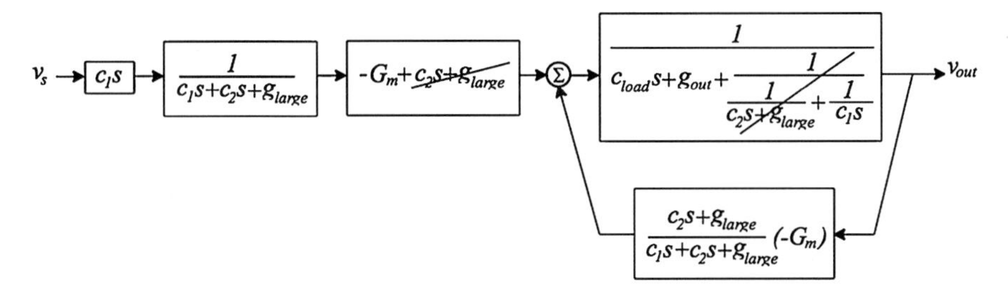

Fig. 6.30 Flow graph for capacitive gain stage in Fig. 6.29

this circuit onto the flow graph in Fig. 6.30 where we will drop the feedback admittance compared to that of the load and output and again compared to the amplifier transconductance. From this flow graph, we extract the transfer function numerator and the loop gain for the system, Eq. (6.28).

$$\begin{aligned}
\text{Num} &= \frac{c_1 s(-G_\text{m})}{c_1 s + c_2 s + g_\text{large}} \frac{1}{c_\text{load} s + g_\text{out}} \\
&= \frac{c_1 s(-G_\text{m}) r_\text{large}}{1 + s(c_1 + c_2) r_\text{large}} \frac{r_\text{out}}{1 + r_\text{out} c_\text{load} s} \\
&= \frac{c_1 s(-G_\text{m}) r_\text{large}}{1 + \dfrac{s}{\omega_1}} \frac{r_\text{out}}{1 + \dfrac{s}{\omega_3}} \\
\text{loop gain} &= \frac{r_\text{out}}{1 + r_\text{out} c_\text{load} s} \frac{(1 + r_\text{large} c_2 s)(-G_\text{m})}{1 + (c_1 + c_2) s r_\text{large}} \\
&= \frac{r_\text{out}}{1 + \dfrac{s}{\omega_3}} \frac{\left(1 + \dfrac{s}{\omega_2}\right)(-G_\text{m})}{1 + \dfrac{s}{\omega_1}} \\
\omega_1 &= \frac{1}{(c_1 + c_2) r_\text{large}};\ \omega_2 = \frac{1}{c_2 r_\text{large}};\ \omega_3 = \frac{1}{c_\text{load} r_\text{out}}
\end{aligned} \tag{6.28}$$

Figure 6.31 shows Bode plots of the transfer function numerator, and the loop gain (dashed trace in plot) responses for the capacitive gain cell using the equations in Eq. (6.28), plus that for the closed-loop transfer function response inferred from these two system functions (plotted as the heavy trace). We see a band-pass response with the mid-band gain set as the ratio of the input capacitance to the feedback capacitance, c_1/c_2; the lower-frequency band edge is defined by the feedback capacitance c_2 and the very large bias resistors r_{large}; and the high-frequency roll-off is defined by the closed-loop gain, the load capacitance, and the OTA transconductance (Fig. 6.32).

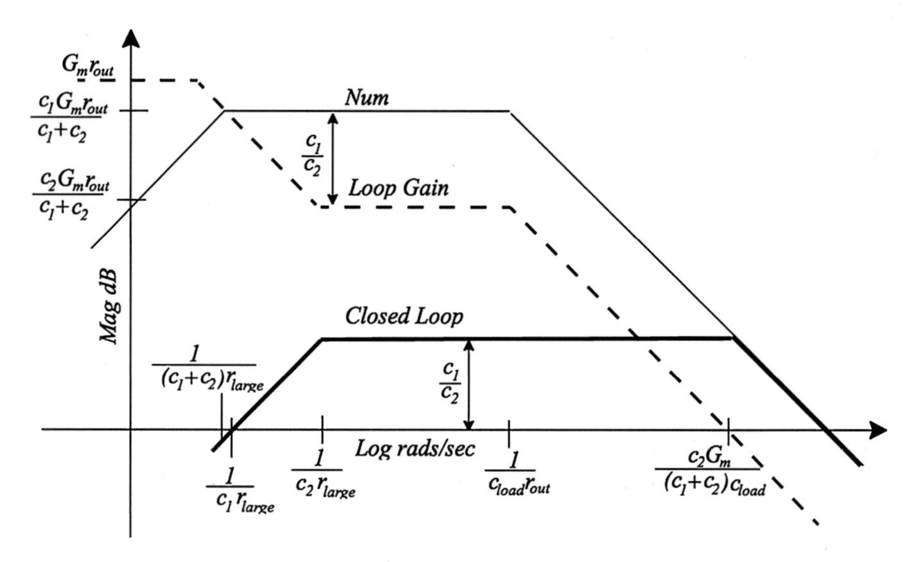

Fig. 6.31 Bode plot for capacitive gain cell showing transfer function numerator, loop gain, and closed-loop responses

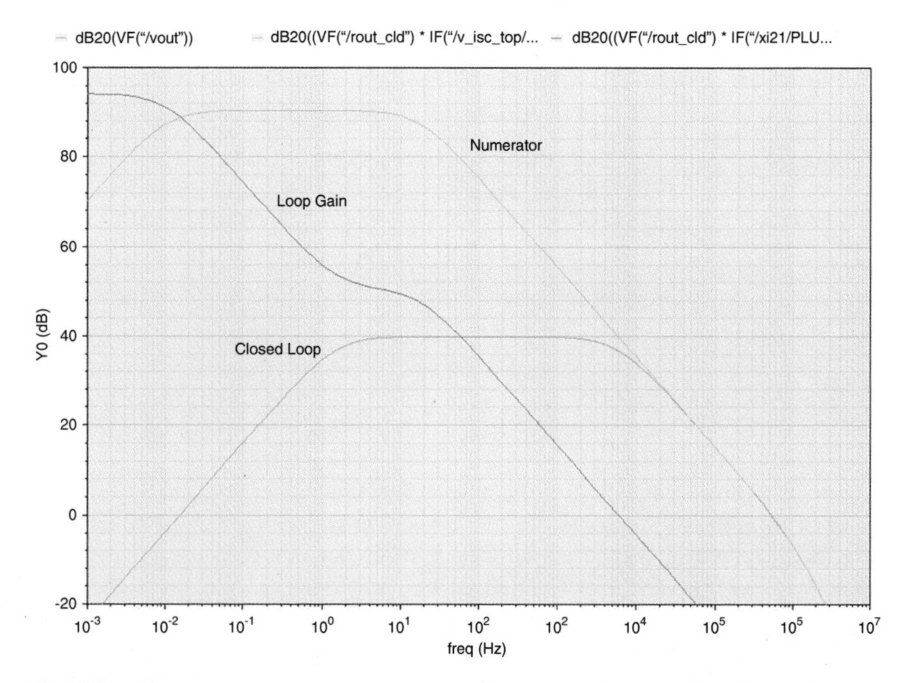

Fig. 6.32 Simulation results for capacitive gain cell showing transfer function numerator, system loop gain, and the close loop responses following the hand calculations shown in Fig. 6.31

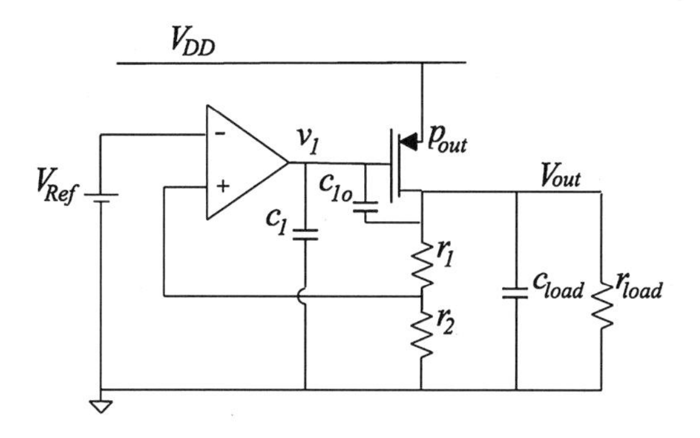

Fig. 6.33 Basic schematic for an LDO regulator

$$f_{low} = 1/(2\pi c_2 r_{large})$$
$$f_{high} = c_2 G_m/(2\pi c_{load}(c_1 + c_2) = G_m/(2\pi c_{load}(1 + A)$$
$$A = c_1/c_2$$

Using a single-stage, folded OTA with cascoded outputs as the amplifier block and circuit parameters $r_{large} = 501\ G\Omega$, $r_{out} = 500\ M\Omega$, $c_1 = 20\ pF$, $c_2 = 0.20\ pF$, results in the simulated responses shown in Fig. 6.33, following well the hand Bode plots in Fig. 6.31. The frequencies of interest in these plots are, the lower and upper bandpass frequencies, and the bandwidt as given below:

ω_2	1.5 Hz × 2π rads Lower BP Corner Frequency
ω_3	18 Hz × 2π rads
$\omega_{\rm bph}$	6 kHz × 2π rads Upper BP Corner Frequency.

The hand analysis results in Eq. (6.28), and the Bode plots in Fig. 6.31 show all the features of the simulations, giving us insight into the operation of the cell and how to change the design elements to meet a particular specification.

Elements of Low Dropout Regulators (LDO's)

A voltage regulator cell is one that takes in a raw voltage that may deliver power over a range of voltage and may also be noisy, to a delivered voltage that is controlled in magnitude, has reduced noise, and shows a low output impedance to its load. A battery, for example, will deliver a voltage that will droop over time from a starting value to an end-of-life value. A system that we would like to power from such a source might only work over a portion of the battery life, or its performance may degrade, depending on the delivered voltage with time. A regulator will deliver a constant voltage over time until closer to the battery end-of-life, maintaining peak performance of its load. A low dropout regulator (LDO) is designed to operate up to a minimum input/output differential voltage to maximize system life.

The circuit in Fig. 6.33 shows the basic elements for an LDO—a gain stage in a feedback loop, a sample of the output voltage that is fed back for comparison with a reference voltage, and an output stage. The load is shown as a capacitance and a shunt resistor where the resistor models the effective power taken by the load and the capacitor provides transient currents and is involved in the compensation of the system. Capacitor c_{1o} models the net gate-drain capacitance comprised of an intrinsic device capacitance and any added circuit capacitance, and c_1 is all other capacitance at node v_1, design, or parasitic. The output voltage is regulated as the supply voltage droops until the output transistor P_{out} goes into triode operation and the loop gain drops in magnitude, losing regulation. The objective is to design the output device to have as low a saturation voltage as possible while meeting area and performance specifications.

The zero-order-starting analysis uses the ideal feedback approximation that the error voltage is very small and can be approximated as zero. Here the positive input to the amplifier becomes equal to the negative input, V_{Ref}, making the output voltage as shown in Eq. (6.29).

$$V_{\text{Ref}} = V_{\text{out}} \frac{r_2}{r_1 + r_2} \tag{6.29}$$

We can see from the schematic in Fig. 6.33 that we have two inversions in the feedback loop so that we may have stability problems. Modeling the amplifier (which may become an OTA, as an input capacitance, c_{in}; a transconductor, G_m; and an output conductance, g_o), we can draw the system flow graph by inspection as shown in Fig. 6.34. The flow graph is drawn in two sections with (a) being the short-circuit current to the output and (b) representing the output node driving point impedance to identify the loop gain.

The short-circuit current transfer relation in Fig. 6.34a is seen to have a pole due to the node at the output of the OTA and a right-half-plane zero defined by the net capacitance across the gate to the drain nodes c_{1o} and the output transistor's

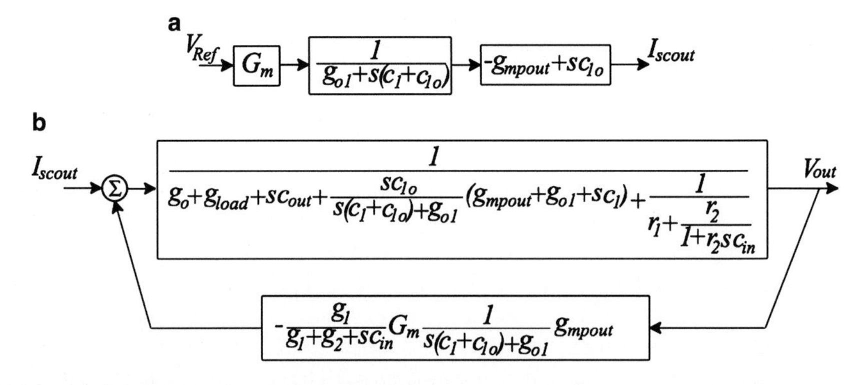

Fig. 6.34 Flow graph for the LDO in Fig. 6.33

transconductance. z_0', the output port impedance with the outer feedback loop zeroed, appears in Fig. 6.34b, upper block, and requires some algebra reduction to discuss further. The design has an inner feedback loop through c_{1o} and the output pMOS transistor, in addition to the outer loop around the OTA. This inner loop is absorbed in the analysis and fully accounted in the flow graph z_0' block. To simplify z_0', we note that the feedback admittance looking into r_1, while the LDO is driving a reasonable load, will be small compared to the load so that we can ignore it. We also drop g_{o1} and sc_1 compared to g_{mpout}, and, combining the output admittances $g_0 + g_{load} = g_{out}$, we write Eq. (6.30) for z_0'.

$$z_0' \approx \frac{1}{g_{\text{out}} + sc_{\text{out}} + \dfrac{sc_{1\text{o}}g_{\text{mout}}}{s(c_{1\text{o}} + c_1) + g_{\text{o1}}}} = \frac{r_{\text{out}}(r_{\text{o1}}s(c_{1\text{o}} + c_1) + 1)}{(r_{\text{o1}}s(c_{1\text{o}} + c_1) + 1)(1 + sc_{\text{out}}r_{\text{out}}) + sc_{1\text{o}}g_{\text{mout}}r_{\text{out}}r_{\text{o1}}} \tag{6.30}$$

In this form, we see z_0' has a second-order denominator producing two poles, plus a zero in the numerator. We also see that capacitance c_{1o}, the capacitance in the inner feedback loop, appears multiplied by $g_{mpout} \times r_{out}$, due to the Miller effect.

The loop gain function, the product of the two blocks in Fig. 6.34b, is seen to have two poles and high low-frequency gain.

A trial design in 0.18 μm CMOS producing 1.7 V from a 1.2 V reference voltage and from a 2 V supply uses a resistor feedback divider with the upper resistor, r_1, set to *500 kΩ* and r_2 set to *1.2 MΩ*, $c_{load} = 50\ pF$, and $r_{load} = 100\ \Omega$. The output drive transistor is *40* fingers of $L = 0.18\ \mu m$, $W = 20\ um$, is used to show the behavior discussed above. Simulation results in Fig. 6.35 show the voltage transfer function numerator as the upper curve, the loop gain function slightly below the first curve, and the closed-loop response, a low-pass filter showing some peaking at the unity gain frequency of the loop gain function.

Figure 6.36 repeats the loop gain magnitude function and adds the phase response. At loop gain unity gain, we see that the phase margin is *49°* consistent with the peaking seen in Fig. 6.35.

We now have analytical expressions for major factors of the LDO transfer relation and can explore specific elements for their effect on the overall response. To improve the phase margin, we can add capacitance to the OTA output node, v_1, as this node for the results above defines the dominant pole. We can also increase c_{1o}, adding Miller capacitance. Figure 6.37 shows the loop gain magnitude, loop gain phase, and the closed-loop response after increase to *1 pF*.

Here we have only looked at the analysis and responses under loaded conditions—we are drawing enough current that the load itself is relatively low, about *100 Ω*, and the current is high enough that the pMOS output transistor's g_{ds} is also high at approximately *13 mS*. Full design will require exercising the design over loads, temperature, supply voltages, and process corners.

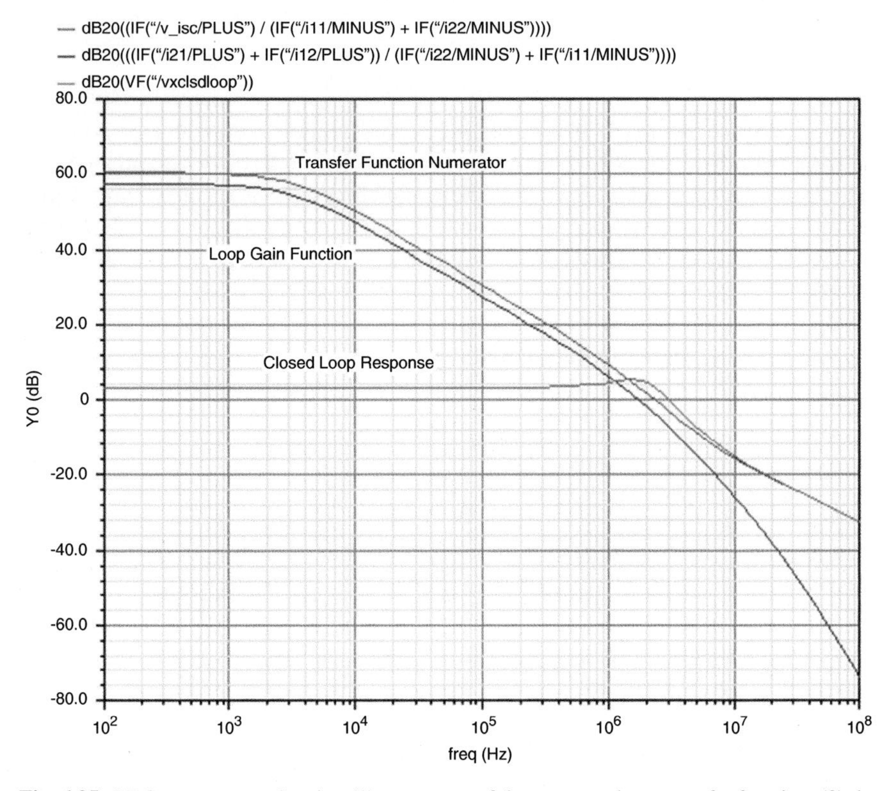

Fig. 6.35 LDO ac response showing (1) numerator of the output voltage transfer function, (2) the LDO loop gain, and (3) the closed-loop response

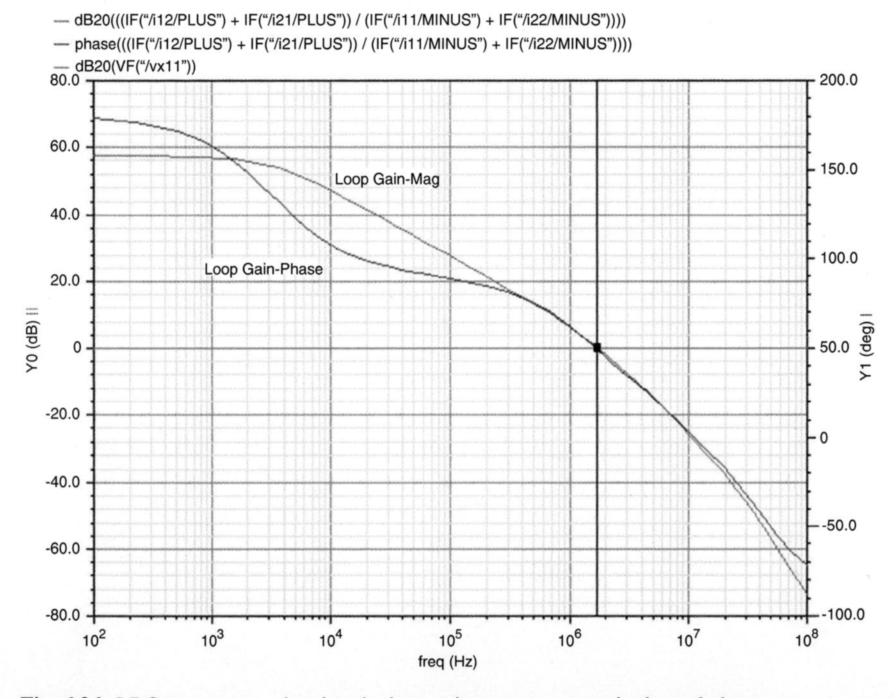

Fig. 6.36 LDO ac response showing the loop gain response, magnitude, and phase

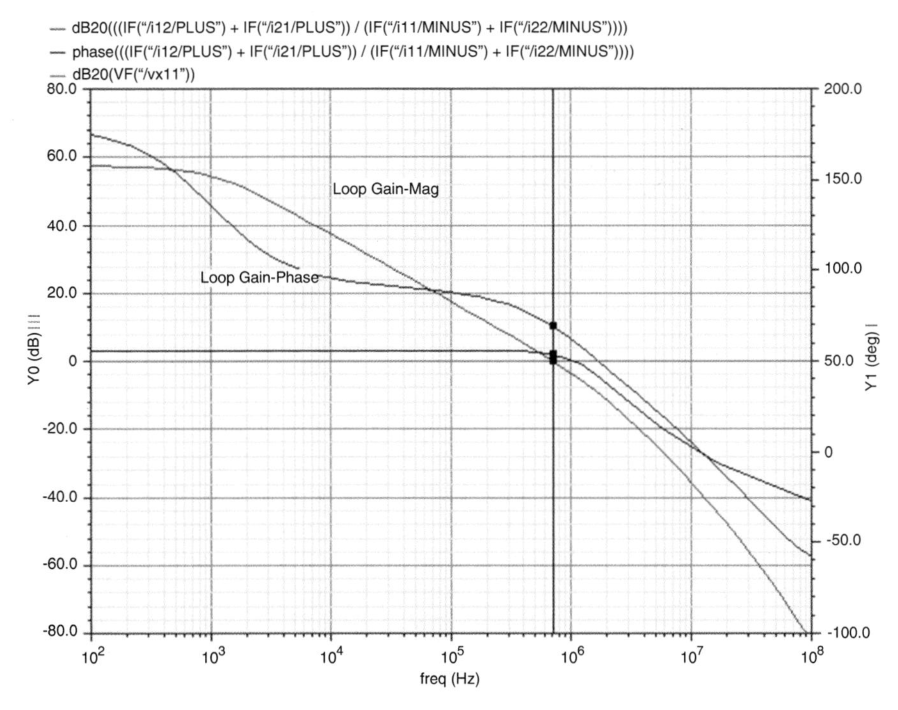

Fig. 6.37 LDO ac response showing the loop gain, magnitude and phase, and closed-loop response with c_{1o}=*1pF*

Elements of Phase-Locked Loop Subsystems

Phase-locked loops are frequency multipliers. A periodic signal is supplied to the input of the PLL where it is compared to a divided down output waveform, Fig. 6.38. Following the zero-order analysis of feedback systems, the inputs are forced equal (here in phase and therefore in frequency) so that we can write for the output frequency f_{out} as in Eq. (6.31) where M is the divider value in the feedback path.

$$f_{\text{in}} = \frac{f_{\text{out}}}{M}$$
$$f_{\text{out}} = M \cdot f_{\text{in}} \tag{6.31}$$

The input and output signals are mostly squared-off sinusoids but may also be sinusoids. Figure 6.39 shows the expanded block diagram for the PLL system as consisting of a phase detector, Ph Det, and charge pump, Ch Pmp, a loop filter H(s), a voltage-controlled oscillator (VCO), and a feedback divider. The property of the signal of interest changes type depending on where in the loop it is considered. The input clock is characterized as a frequency f_{in}, but its phase is the focus in the phase

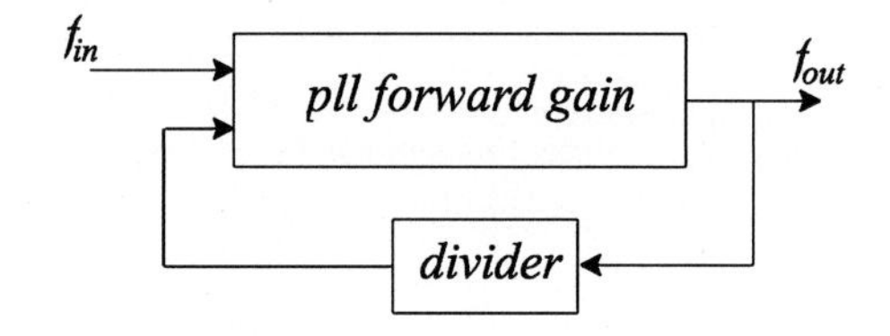

Fig. 6.38 Phase lock loop simple block diagram

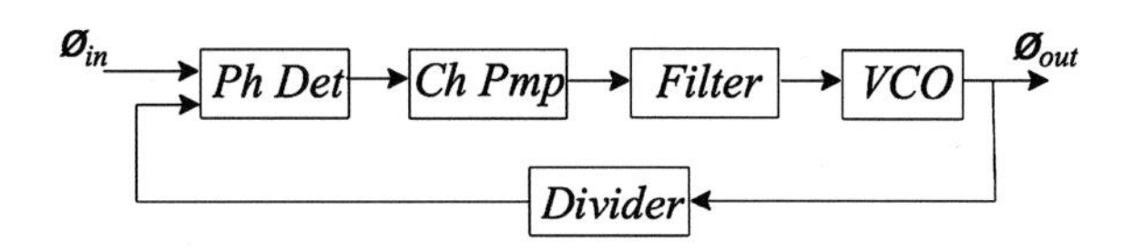

Fig. 6.39 Phase lock loop block diagram

detector block so we switch input variables to indicate this fact. The output of the phase detector is a sequence of pulses with width equal to the phase difference between the input reference signal ϕ_{in} and the divided feedback clock ϕ_{out}/M. Assuming squared-off signals, the digital phase detector generates a signed pulse on the rising edge of the first arriving input signal and terminates the pulse on the rising edge of the other input signal. A positive pulse is initiated if the input reference signal arrives first, a negative if the feedback signal arrives first. This signed phase pulse, now in the voltage-time domain, directs a charge pump to inject current for the duration of the pulse, into (positive) or out of (negative) a loop filter. The filter converts the current pulses from the charge pump, whose integral is the net corrective charge signal for the loop, into an average, filtered voltage.

The voltage output of the filter controls the frequency of the voltage-controlled oscillator block. The slow varying filtered voltage signal speeds up (the VCO frequency increases) or slows down the VCO to maintain a lock on the output frequency of the oscillator. The output of the VCO is squared up to digital levels to be used by the system as a clocking signal, and it is returned through a divider block to the phase detector completing the loop.

If the input reference signal at the phase detector arrives before the divider signal, this is an indication that the PLL output is lagging the reference signal. Assuming a positive K_{vco}, a positive charge injection into the loop filter will raise the filter voltage to speed up the VCO, reducing the phase difference between inputs. An early divided input signal indicates that the VCO is fast and should be slowed down by reducing the controlling filter voltage. A negative charge pulse is injected into the filter for this case.

The accuracy of the PLL output frequency is directly related to the accuracy of the input frequency. The input reference frequency is generally supplied from a crystal oscillator having very precise operating frequency due to the use of a very high Q crystal resonator driven in a feedback configuration.

Analysis of the PLL system is simplified as the blocks are mainly unidirectional and reasonably free of loading effects. The phase detector output is a pulse whose width represents the fraction of a reference cycle that the charge pump current is to be directed onto the loop filter. The phase detector converts the input phase difference between the reference input and the divided output to this fraction (Eq. 6.32), defining the Ph Det gain factor, K_{PD}.

$$\begin{aligned}\mathrm{Pulse}_{\mathrm{PD,out}} &= \frac{\Delta t_{\mathrm{out}}}{T_{\mathrm{ref}}} = \frac{\Delta\phi_{\mathrm{in}}}{2\pi} \\ K_{\mathrm{PD}} &= \frac{1}{2\pi}\end{aligned} \tag{6.32}$$

The charge pump model is simply a current I_{ChPmp}. It is multiplied by the phase detector output pulse and injected into the filter. The filter transfer function is its impedance. The filter voltage controls the frequency of the VCO.

The VCO converts a voltage to a frequency so that it will add a factor K_{VCO} to the signal path converting the filter voltage to a frequency. Its output needs to be *phase* in order to match the input to the phase detector. Since phase is the integral of frequency in radians, the VCO transfer function in Laplace space becomes as shown in Eq. (6.33). The 2π factor changes the units from Hz to radians/s.

The divider simply divides the VCO's output phase by M so that the PLL loop becomes as shown in Fig. 6.40.

$$H_{\mathrm{VCO}} = \frac{K_{\mathrm{VCO}} 2\pi}{s} \tag{6.33}$$

The magnitudes, other than that for the phase detector, are at this point unspecified. The loop filter in particular needs attention. The purpose of the filter is to average out pulses generated by the phase detector and charge pump and to compensate the loop to insure stability. Consider first using a capacitor as the filter with $z(s) = 1/cs$. The loop gain, Eq. (6.34), has two poles at zero Hz, making the loop unstable. We can compensate the loop by adding a zero in the loop filter to add phase:

$$LG_{\mathrm{PLL}} = \frac{-1}{2\pi} I_{cp} \frac{1}{cs} \frac{K_{\mathrm{VCO}} 2\pi}{s} \frac{1}{M} \tag{6.34}$$

Consider the filter in Fig. 6.41. The impedance for this filter is given in Eq. (6.35).

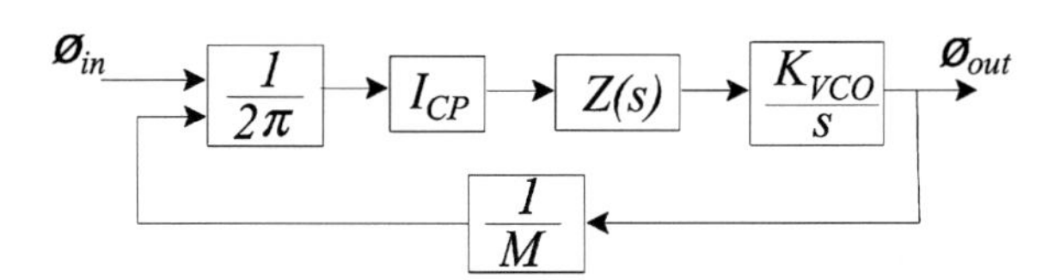

Fig. 6.40 Block flow graph for the PLL system showing first-order block transfer relations

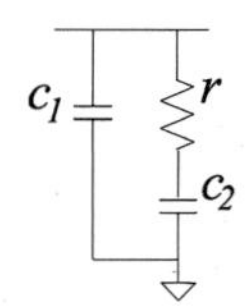

Fig. 6.41 Loop filter with added zero and an additional pole

$$
\begin{aligned}
Z_{\text{filt}}(s) &= \frac{1}{c_1 s + \dfrac{1}{r + \dfrac{1}{c_2 s}}} \\
&= \frac{1 + r c_2 s}{(c_1 + c_2) s \left(1 + \dfrac{r c_1 c_2 s}{c_1 + c_2}\right)}
\end{aligned}
\tag{6.35}
$$

The resulting loop gain can now be written as in Eq. (6.36) where we have added a zero and a pole.

$$
\begin{aligned}
LG_{\text{PLL}} &= \frac{-1}{2\pi} I_{\text{CP}} \frac{1 + r c_2 s}{(c_1 + c_2) s \left(1 + \dfrac{r c_1 c_2 s}{c_1 + c_2}\right)} \frac{K_{\text{VCO}} 2\pi}{s} \frac{1}{M} \\
&= \frac{-I_{\text{CP}} \left(1 + \dfrac{s}{\omega_z}\right) K_{\text{VCO}}}{M s^2 (c_1 + c_2) \left(1 + \dfrac{s}{\omega_{\text{p}}}\right)} \\
\omega_{\text{z}} &= \frac{1}{r c_2}; \ \omega_{\text{p}} = \frac{1}{\dfrac{r c_1 c_2}{c_1 + c_2}}
\end{aligned}
\tag{6.36}
$$

We see, from Eq. (6.36), that the zero will always be at a lower frequency than that of the pole added by the *r*–*c* branch. We need to position the loop gain function such that we cross unity gain somewhere between the zero and the pole frequencies, ω_z and $\omega_{p.}$ We can set the separation between these break frequencies to improve the phase margin. A Bode plot for a potential loop gain function is shown in Fig. 6.42. At low frequency, we have a two-pole response until we reach the zero where we change to a 20 dB/decade roll-off, one pole response region. The design challenge is now clear—position the zero and pole to yield the greatest phase margin. The situation shown here will lead to poor phase margin as the unity gain point of the loop gain function is close to the pole frequency f_p where the phase of the loop gain function will be dropping decreasing phase margin.

In order to get good filtering, to be able to average out the charge pump pulses to give the VCO a quiet signal, the unity gain crossing of the loop gain function should be at least 10–20 times lower than the input frequency. This gives us a target for

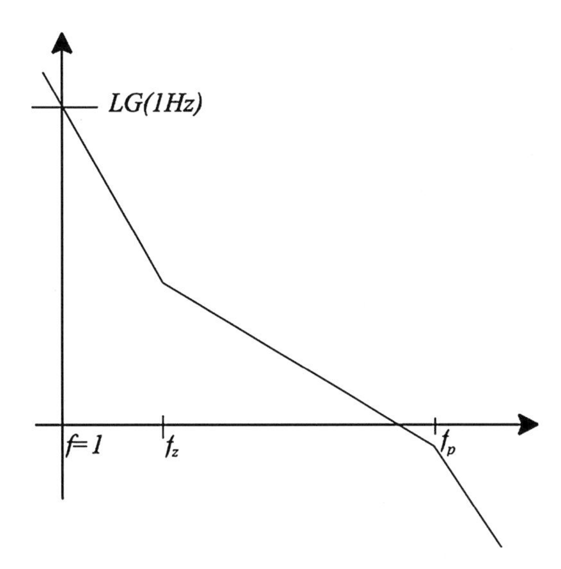

Fig. 6.42 Loop gain with *c–r/c* filter in Fig. 6.41

closed-loop bandwidth defined by the unity gain crossing of the loop gain function. The lower the bandwidth, the better the filtering but also the greater the settling time– tradeoff time.

Symmetry suggests we position the zero and pole equal distance geometrically from the unity crossing. This will give the maximum phase margin for a given f_p/f_z frequency ratio, R.

Starting with the input reference frequency, we set f_{unity} for the loop gain to $\sim f_{ref}/20$, giving us a better filtering frequency than we would get by using a divider of *10*.

To place f_{unity} at the geometric center between f_p and f_z, we place these a factor $\sqrt{R}$ above f_{unity} for the pole and $\sqrt{R}$ below f_{unity} for the zero, where $R = f_p/f_z$, as shown in Fig. 6.43.

The loop gain can now be written as in Eq. (6.37), using LG(1 Hz), the loop gain magnitude at 1 Hz.

$$LG_{\mathrm{PLL}}(s) = \frac{-I_{\mathrm{CP}}K_{\mathrm{VCO}}}{M(c_1 + c_2)} \frac{\left(1 + \frac{s}{\omega_{\mathrm{z}}}\right)}{s^2\left(1 + \frac{s}{\omega_{\mathrm{p}}}\right)}$$
$$LG_{\mathrm{PLL}}(1\,\mathrm{Hz})_{s=2\pi} = \frac{-I_{\mathrm{CP}}K_{\mathrm{VCO}}}{M(c_1 + c_2)(2\pi)^2} \tag{6.37}$$

We have the input reference frequency, f_{ref}, and the desired output frequency, f_{PLL}, from the requirements for the PLL. With the output frequency specified, we can design the VCO and characterize it for the VCO gain, K_{VCO}.

Next we move on to the filter. We define the loop gain unity gain crossing f_{unity} to be $f_{\mathrm{unity}} \approx f_{ref}/20$. We select R, the ratio of the pole frequency to the zero, $R = f_p/f_z$. From Eq. (6.37) we find that phase margin is given as Eq. (6.38).

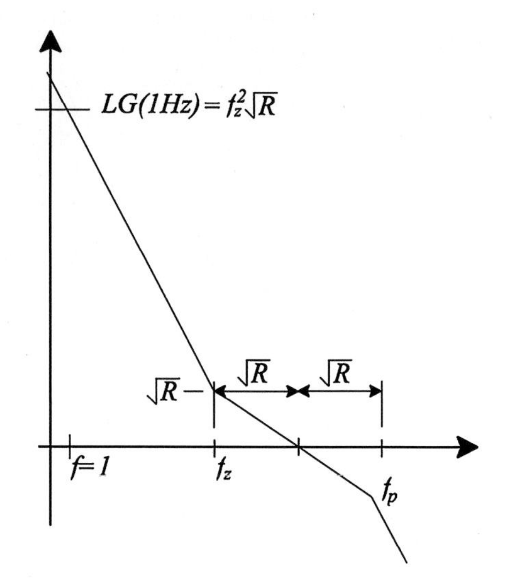

Fig. 6.43 Balanced loop gain for maximum phase margin for a given pole/zero ratio R

$$
\begin{aligned}
PM &= \text{Angle}\left(\frac{I_{CP}K_{VCO}}{M(c_1+c_2)}\frac{-\left(1+\frac{s}{2\pi f_z}\right)}{s^2\left(1+\frac{s}{2\pi f_p}\right)}\right)_{s=j2\pi f_{\text{unity}}} \\
PM &= \text{atn}^{-1}\left(1+\frac{s}{2\pi f_z}\right) - \text{atn}^{-1}\left(1+\frac{s}{2\pi f_p}\right)_{s=j2\pi f_{\text{unity}}} \\
PM &= \text{atn}^{-1}\left(1+j\sqrt{R}\right) - \text{atn}^{-1}\left(1+\frac{j}{\sqrt{R}}\right) \\
R &= \frac{f_p}{f_z}
\end{aligned}
\tag{6.38}
$$

From Eq. (6.38), we calculate that for $R=6$, $PM=45°$, and $R=10$ and $R=15$, we get $PM=55°$ and $60°$ respectively. A pole-zero ratio of 6 provides good phase margin while keeping the upper pole low. Keeping this upper pole low helps with the phase noise at the output.

We now have all but one constraint to the design of the filter. We need to select one of the capacitors, at this point arbitrarily, but it should be sized with consideration to noise, again to keep the phase noise of the PLL output as low as possible.

The loop gain zero frequency is given in Eq. (6.39). Arbitrarily selecting c_2, we solve for the resistor r in terms of c_2 and f_z.

$$
\begin{aligned}
f_z &= \frac{1}{2\pi r c_2} \\
r &= \frac{1}{2\pi f_z c_2}
\end{aligned}
\tag{6.39}
$$

Given the pole/zero ratio R and f_z, we can obtain c_1 as given in Eq. (6.40).

$$\begin{aligned} R &= \frac{f_p}{f_z} \\ &= 1 + \frac{c_2}{c_1} \\ c_1 &= \frac{c_2}{R-1} \end{aligned} \tag{6.40}$$

We need the charge pump current. Since we are crossing f_{unity} with a slope of *20 dB/decade*, the magnitude of the loop gain function at the zero frequency f_z is $\sqrt{R}$. From here going to lower frequencies, we rise at *40 dB/decade*, making the loop gain function magnitude at *1 Hz*, $LG(1\ Hz) = f_z^2\sqrt{R} = f_z \times BW$. Using this result with Eq. (6.37), we find the charge pump current in Eq. (6.41).

$$\begin{aligned} LG_{\mathrm{PLL}}(1\,\mathrm{Hz})_{s=2\pi} &= \frac{I_{\mathrm{CP}}K_{\mathrm{VCO}}}{M(2\pi)^2(c_1+c_2)(2\pi)^2} = f_z^2\sqrt{R} \\ I_{\mathrm{CP}} &= \frac{f_z^2\sqrt{R}M(2\pi)^2(c_1+c_2)}{K_{\mathrm{VCO}}} \end{aligned} \tag{6.41}$$

A design process will be as follows:

1. Knowing the input reference frequency, select the desired bandwidth.
2. Select the pole/zero ratio R. Calculate f_z and f_p.
3. Select c_2 arbitrarily or better, from a noise analysis requirement.
4. Given c_2 and f_z, calculate the filter resistor r.
5. From f_z, R, and c_2, calculate c_1.
6. Given BW, K_{VCO}, and f_z, find the magnitude of $LG(1\ Hz) = f_z^2\sqrt{R} = f_z \times BW$. Use this to calculate the charge pump current.

For a sample design, using $f_{ref} = 1\ MHz$, and loop divider $M = 100$, we select a bandwidth of *50 kHz* and a pole/zero ratio $R = 16$. From these, we calculate $f_{ref} = 12.5\ kHz$ and $f_p = 200\ kHz$ and $LG_{PLL}(1\ \mathrm{Hz}) = 6.25\ E8$. The loop gain function can be written as Eq. (6.42). In this form, simulate using MatLab, XCEL, or SPICE tools to obtain the responses shown in Fig. 6.44 for the loop gain function, magnitude and phase, and the closed-loop response PLL response. The bandwidth of the PLL is *50 kHz*, and the phase margin is *62°* due to the selection of $R = 16$ and pole-zero positioning. Closed-loop gain is *40 dB* and well behaved at the bandwidth edge.

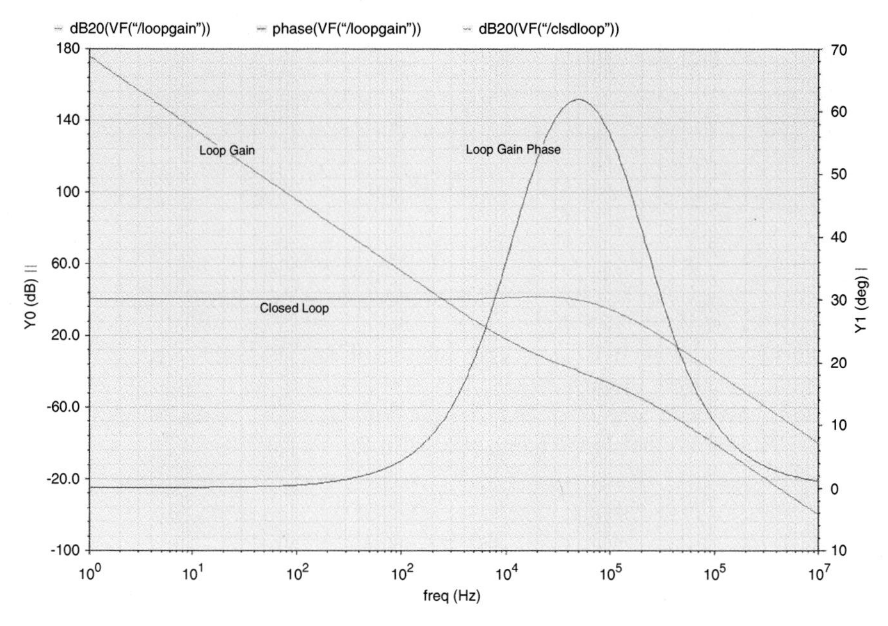

Fig. 6.44 Simulated loop gain, magnitude and phase, and closed-loop response using Eq. (6.42)

$$\begin{aligned} LG_{\mathrm{PLL}}(s)_{s=2\pi} &= LG_{\mathrm{PLL}}(1\mathrm{Hz})(2\pi)^2 \frac{\left(1+\frac{s}{2\pi f_{\mathrm{z}}}\right)}{s^2\left(1+\frac{s}{2\pi f_{\mathrm{p}}}\right)} \\ &= f_{\mathrm{z}}\mathrm{BW}(2\pi)^2 \frac{\left(1+\frac{s}{2\pi f_{\mathrm{z}}}\right)}{s^2\left(1+\frac{s}{2\pi f_{\mathrm{p}}}\right)} \end{aligned} \tag{6.42}$$

Table 6.2 shows numerical results for this PLL design using the VCO gain factor $K_{\mathrm{vco}} = 10^7$ Hz/V.

The process for designing a phase lock loop presented has the advantage of direct control of the interaction of elements. Only the loop filter particulars were required to fully define the loop and verify closed-loop behavior. To give the reader a better feeling for the whole subsystem, a set of example cells for each of the blocks will be briefly described.

A digital phase detector is composed of two D-type flip-flops and a reset path. We wire each flip-flop as a 'trigger on clock' cell with one monitoring the reference clock and the other the divided output clock. From a reset condition, the flop that first receives a rising edge on its clock input gets set and directs the following charge pump current source to inject current into or out of the filter, depending on

Table 6.2 PLL design parameters

Parameter	Value	Comment
f_{ref}	1E6	Hz, Spec
M	100	Given, Spec
Bandwidth	50	kHz, selected
R	16	Pole/zero ratio, selected
K_{vco}	10^7	Hz/Volt, VCO characterization
f_z	12.5	KHz, calculated, $f_z = BW/\sqrt{R}$
f_p	200	kHz, calculated, $f_p = f_p \times R$
c_2	20	pF, selected
c_1	1.333	$pF,\ c_1 = \frac{c_2}{R-1}$
r	637	K Ohms, calculated
$G(1\ Hz)$	6.25E8	$G(1) = f_z \times BW$
Phase margin	62	Degrees, calculated, sim
I_{cp}	5.26	μA, calculated

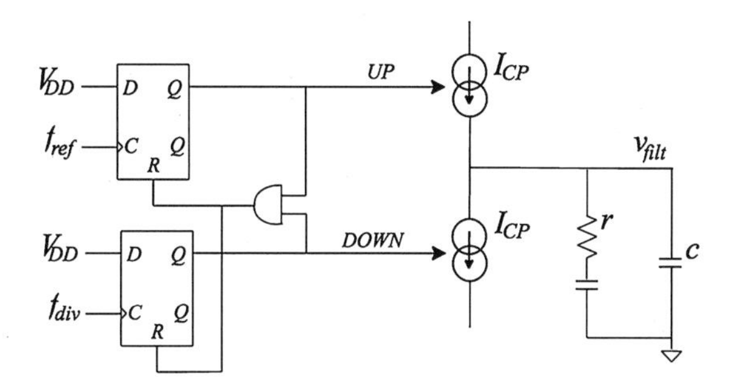

Fig. 6.45 Abbreviated schematic of a phase detector, charge pump, and loop filter

which flip-flop gets set. The charge pump delivers a current to the filter until the other signal arrives, setting its flip-flop high. The state of having received both rising edges initiates a reset of the phase detector that also shuts off the charge pump. This operation can be accomplished using the functional schematic shown in Fig. 6.45 where the upper flip-flop generates the "up" signal to raise the filter voltage on receiving the reference clock before the divided clock by switching in the positive current source. On the divided feedback clock arriving first, the "down" signal is generated and the filter voltage receives a negative current pulse reducing v_{filt}.

The VCO block can be implemented in many ways, two of which are shown in Fig. 6.46. The VCO type in (a) is an *l–c* harmonic oscillator with varactor frequency control and the one in (b) is a ring oscillator with bias current control.

Finally, the loop divider block is simply a digital counter set to count M pulses from the VCO, generate an output pulse, and restart the count.

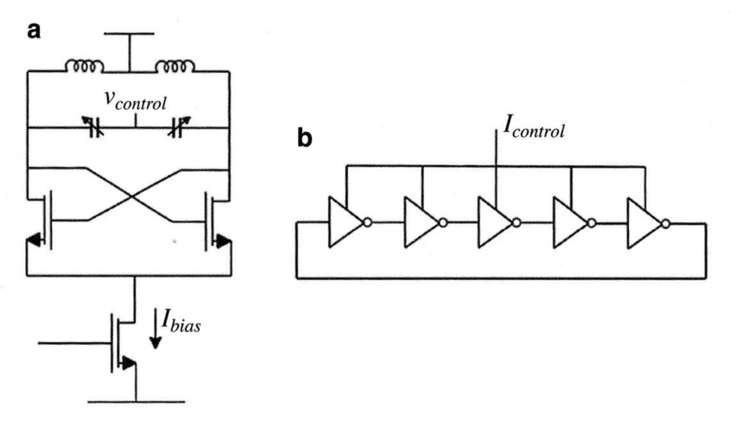

Fig. 6.46 Two VCOs, (**a**) a harmonic oscillator with varactor control and (**b**) a ring oscillator with current bias control

Elements of Switching Regulators

Switching regulators generally have higher conversion efficiency than series regulators such as the LDO considered earlier in this chapter. An LDO operates by modulating the voltage across the series control element (e.g., pMOS transistor) such that there is a voltage ($V_{\text{in}} - V_{\text{out}}$) across this element. Since the load current flows through this control element, the regulator dissipates $I_{\text{load}} \times (V_{\text{in}} - V_{\text{out}})$ power in addition to the power in the rest of the cell. A switching regulator in contrast switches the input power between the unregulated voltage source and ground through low impedance devices so that the net dissipated power in the control elements is small, followed by a filter to deliver a constant voltate to the load. The schematic in Fig. 6.47 shows this operation. The input to the inductor is alternately attached to V_{in} and ground through switches operating on opposite phases of clock ϕ.

The pulse-width modulator (PWM) block converts the input voltage V_{in} into a periodic clock signal with magnitude V_{in} and duty cycle D, to drive the filter and load. The low-pass filter delivers an averaged voltage onto the load. Using reactive elements in the filter and low loss switching minimizes power losses in the regulator block. The capacitor's equivalent series resistance (ESR) is included in the schematic. The clock is required to be higher than the bandwidth of the filter to reduce output ripple.

The dynamics of the system can be improved by adding feedback as shown in Fig. 6.48. As with the PLL phase detector, the PWM changes the signal from the continuous domain, D_{cont}, to an equivalent duty-cycle domain, D. The signal information does not change, only its representation. The voltage delivered to the filter is given as the product of the duty cycle, D, and the input voltage V_{in}. Increasing V_{in} increases the effect of the PWM on the output proportionately.

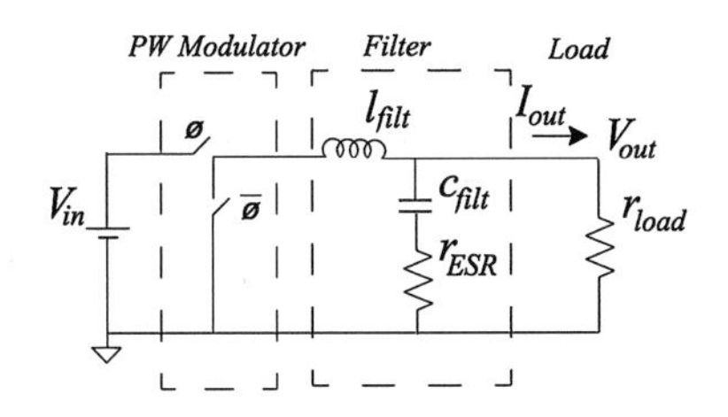

Fig. 6.47 Basic switching regulator schematic

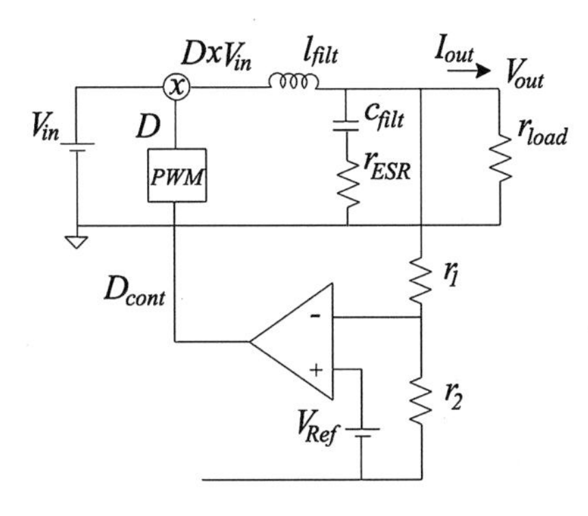

Fig. 6.48 Switching mode power supply core with added feedback path

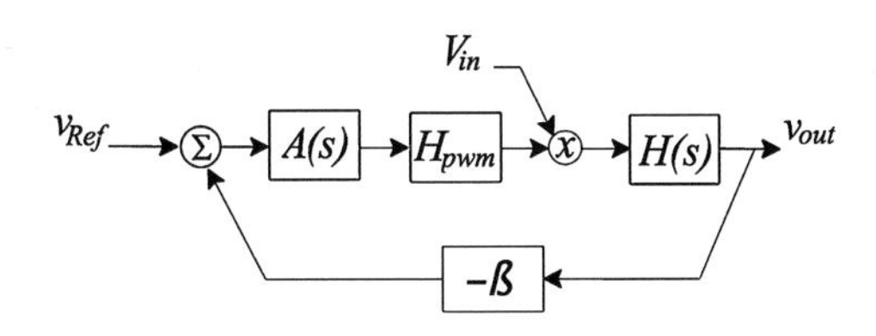

Fig. 6.49 Flow graph for switching mode regulator

Table 6.3 Specification, for example, switching regulator

Parameter	Value	Comment
V_{in}	10	Volts
V_{out}	5	Volts
I_{load}	2	Amps-peak
V_{ref}	1.2	Volts
F_{switch}	200	kHz

The system flow graph is shown in Fig. 6.49 where $A(s)$ is the voltage gain of the amplifier and H_{pwm} and the transfer function for the *PWM*, is *1*. β is the voltage divider composed of r_1 and r_2 used to set the DC gain from V_{ref} to V_{out}.

A partial specification for a switching regulator is shown in Table 6.3.

In steady-state operation, the current through the inductor is triangular with peak-to-peak value greatest at $D = 0.5$. Making this maximum deviation ~*1*/10 of the specification load current, we can solve for the inductor as in Eq. (6.43).

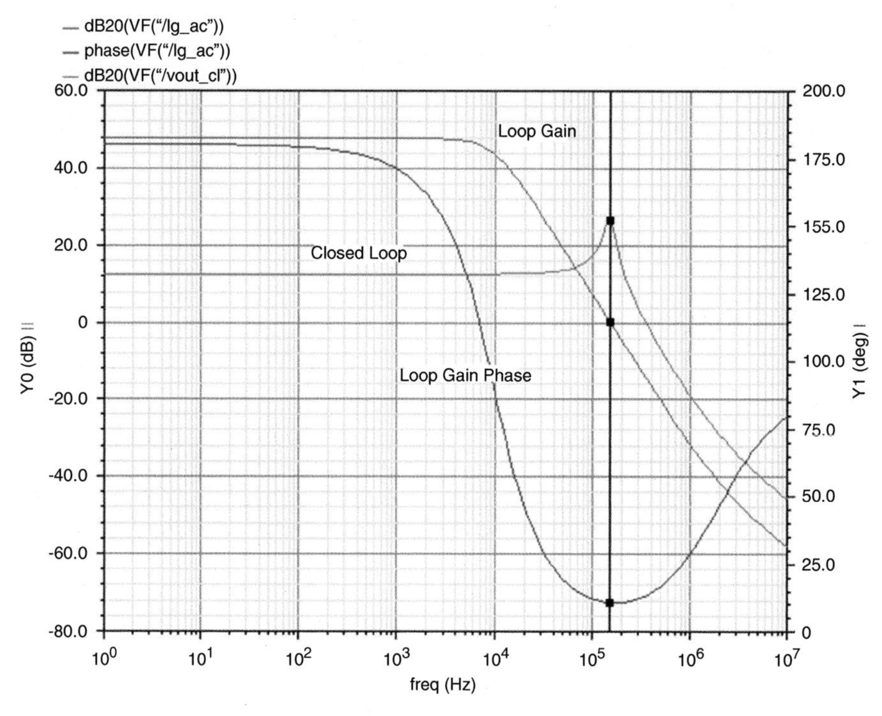

Fig. 6.50 Loop gain magnitude and phase, and closed-loop response for the uncompensated switching regulator

$$
\begin{aligned}
V_{\text{ind}}(t) &= \frac{L\Delta I_{\text{ind}}}{\Delta T} \\
L &= \frac{(V_{\text{in}} - V_{\text{out}})\frac{T}{2}}{\frac{I_{\text{load}}}{10}} \\
&= 62.5\,u\text{H}
\end{aligned}
\tag{6.43}
$$

The LC resonant frequency should be less than ~$F_{switch}/10$. Setting it to $1/20$th, we will use $10\ kHz$, making the $c_{filt}4.05\ \mu F$. r_{load} is $2.5\ \Omega$ from the specifications of maximum current and output voltage. The voltage divider resistors are set to $r_1 = 152$ and $48\ k\Omega$, and the amplifier gain is 100. Simulations show a phase margin of $10°$, Fig. 6.50. The loop gain phase function is seen to drop quickly near the double poles at the LC resonant frequency and then rise due to the ESR zero. The low-frequency loop gain is the product of the amplifier gain (100), the voltage divider factor (0.24), and the $V_{in}(=10)$, $47.6\ dB$.

To improve the phase margin, we can lower the frequency of the ESR zero, but this will increase the output voltage ripple as the capacitor is charged and discharged every cycle. Instead we can add compensation to the loop. First, we

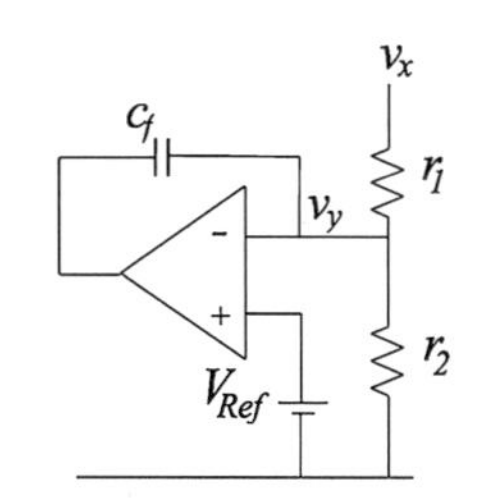

Fig. 6.51 Loop compensation—adding a pole to the loop gain function

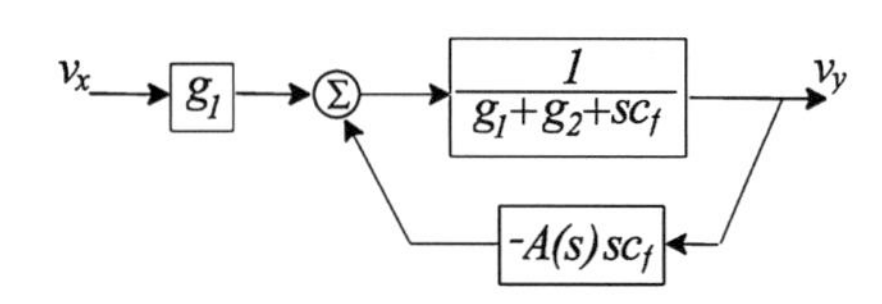

Fig. 6.52 Flow graph for the schematic in Fig. 6.51 for adding a pole to the loop gain function

see that the closed-loop bandwidth is approaching the switching frequency F_{SW} due to the low-frequency gain (LFGain). By adding a low-frequency pole, we can shape the loop gain function to place its unity gain crossing at the desired bandwidth such as $BW = 10\,kHz$. To achieve this bandwidth, we place this low-frequency dominant pole at $f_{\text{pole}} = BW/LFGain = 10\,kHz/240 = 41.7\,Hz$. Such a low frequency will require a very large capacitor. Since we have an amplifier in the feedback loop, we can use this in an active filter. Consider adding a feedback capacitor as shown in Fig. 6.51. The drive comes from the regulator output, and its load is the high-z, pulse-width modulator cell.

The flow graph for adding pole compensation into the loop (feedback Z-form) from the input signal x to the amplifier negative input, y, is shown in Fig. 6.52. An ideal op amp is assumed with amplifier gain $A(s)$.

Solving for the transfer function $v_y(s)/v_x(s)$, we get the single-pole relation in Eq. (6.44), showing the Miller multiplication of c_f. Placing this pole at $41.7\,Hz$ requires that $c_f = 1.036\,nF$. The impact of this pole is seen in the simulation responses in Fig. 6.53. The loop gain function is now seen to have a low-frequency pole and a unity crossing near 10 kHz. Phase margin is still poor at ~$17°$, evident in the peaking in the closed-loop response:

$$\frac{v_y(s)}{v_y(s)} = \frac{\dfrac{g_1}{g_1+g_2+sc_f}}{1+\dfrac{1}{g_1+g_2+sc_f}A(s)sc_f} = \frac{\dfrac{r_2}{r_1+r_2}}{1+\dfrac{sc_f(1+A(s))}{g_1+g_2}} = \frac{\alpha}{1+sc_f(1+A(s))r_1||r_2} \tag{6.44}$$

To gain phase, we could add a zero to the loop gain function. We see from Fig. 6.53 that we transition from a single-pole to a triple-pole response at about $10\,kHz$. Adding a zero at this frequency or below will improve the phase response. Adding a

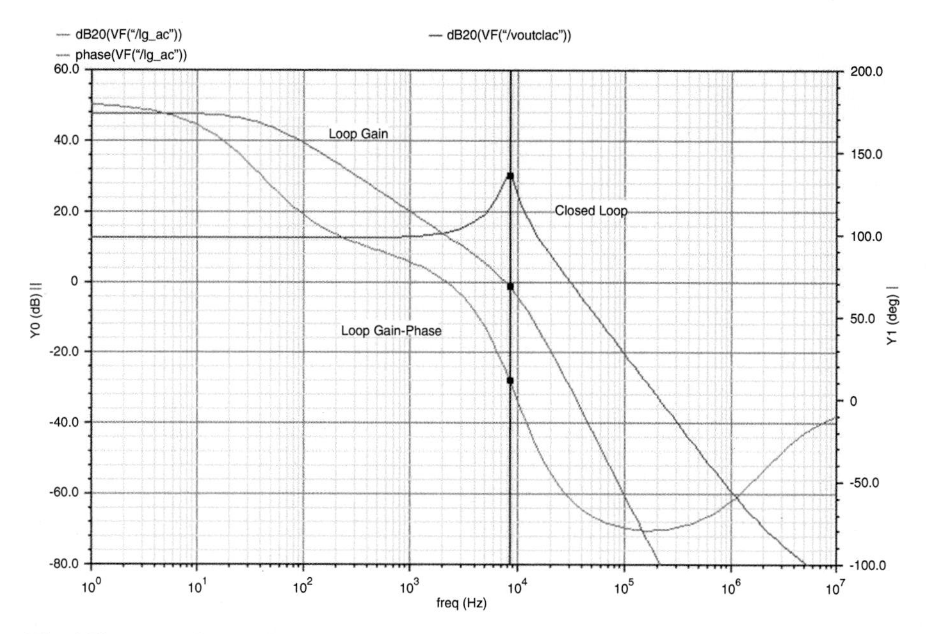

Fig. 6.53 Loop gain magnitude and phase, and closed-loop response for dominant pole compensation of the switching regulator

capacitor across r_l in the schematic in Fig. 6.51 will add a zero to the loop gain response. We can reuse the flow graph in Fig. 6.52, substituting $y_l = sc_l + g_l$ for g_l. The modified transfer relation becomes as shown in Eq. (6.45).

$$\frac{v_y(s)}{v_y(s)} = \frac{\frac{y_1}{y_1+g_2+sc_f}}{1+\frac{1}{y_1+g_2+sc_f}A(s)sc_f} = \frac{\frac{g_1+sc_1}{g_1+sc_1+g_2+sc_f}}{1+\frac{1}{g_1+sc_1+g_2+sc_f}A(s)sc_f}$$

$$= \frac{\frac{r_2}{r_1+r_2}(1+sc_1r_1)}{1+r_1||r_2(sc_1+sc_f(1+A(s))} \tag{6.45}$$

Placing the zero at *10 kHz* requires a capacitor c_l of *104 pF*. We can use this to recalculate c_f for the pole location, but we are adding *0.1 nF* to *101* × *1.036 nF* (due to the Miller multiplier), so we ignore the perturbation in the pole location. The addition of this zero results in the loop gain and closed-loop responses shown in Fig. 6.54.

Phase margin is now ~*80*°. The closed-loop response shows broadened peaking. More tailoring of the response can be done—move the zero to lower frequencies, for example, or add another *r–c* feedback branch across the amplifier. As with the LDO, additional load conditions must be considered as well as variations in input voltage and amplifier gain as we saw these all affect the system response and stability.

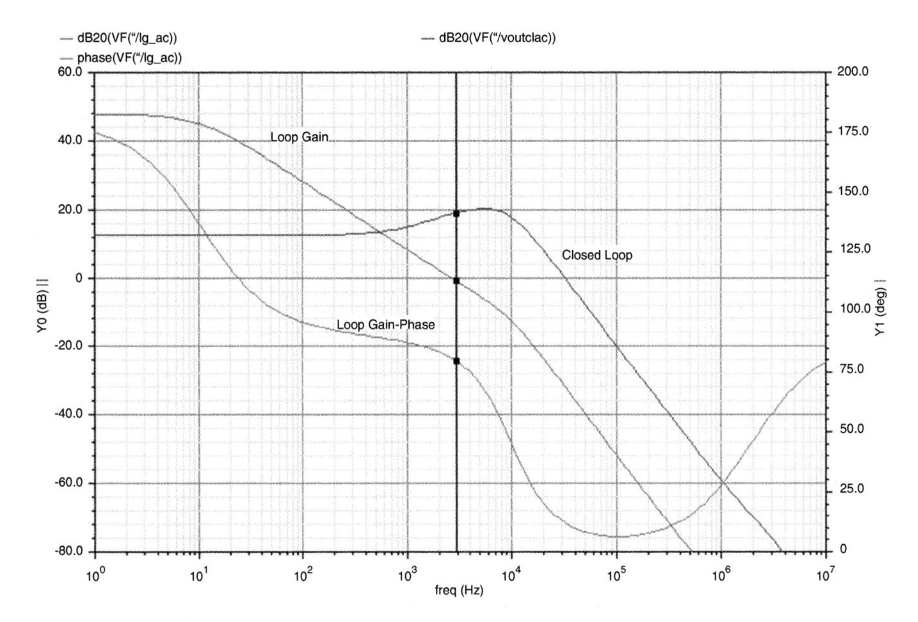

Fig. 6.54 Loop gain magnitude and phase, and closed-loop response for dominant pole compensation and added zero

Elements of G_m–C Filters

Active filters utilize gain stages with passive elements to achieve various transfer functions. We have already seen this type of design in the compensation of the switching regulator where we added a capacitor across an op amp to obtain a Miller-multiplied effective capacitance at the amplifier input impedance. Here we consider using transconductance stages, OTAs, as the active element. Consider the circuits in Fig. 6.55. These are simple G_m–C filters composed of OTAs and capacitors. Their transfer functions are given in Eq. (6.46) as an integrator for (a) and as a low pass for (b) and (c) as two realizations for the same transfer function:

$$H_{\mathrm{a}}(s) = \frac{G_{\mathrm{m}}}{sc_1} = \frac{1}{\frac{sc_1}{G_{\mathrm{m}}}}$$

$$H_{\mathrm{b}}(s) = \frac{G_{\mathrm{m}}\frac{1}{sc_1}}{1 + G_{\mathrm{m}}\frac{1}{sc_1}} = \frac{1}{1 + \frac{sc_1}{G_{\mathrm{m}}}} \tag{6.46}$$

It is clear how to generate the integrator function from these components and, while the low-pass form can also be generated intuitively, let us proceed more systematically. Figure 6.56 shows the flow graph for the low-pass G_m–C filter in Fig. 6.55b,

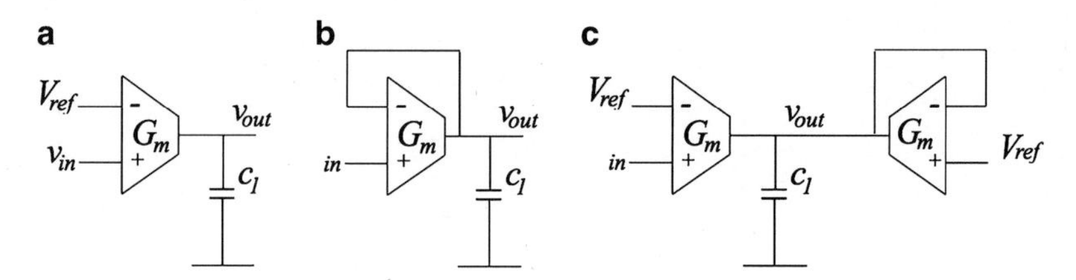

Fig. 6.55 Simple G_m–C filters, (**a**) integrating stage, (**b**) low-pass stage, and (**c**) alternative low-pass stage

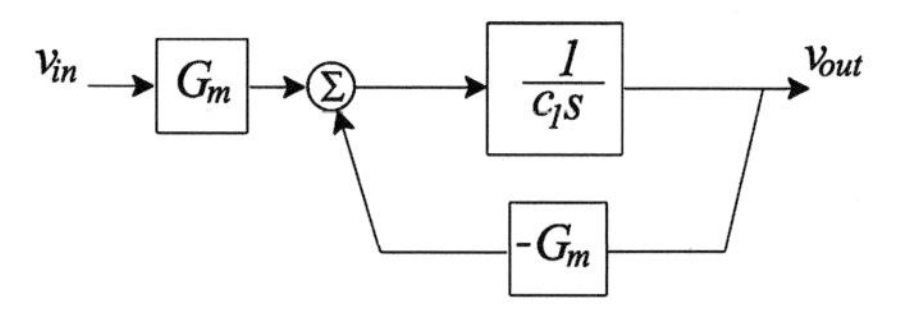

Fig. 6.56 Flow graph for low-pass G_m–C filter in Fig. 6.55b

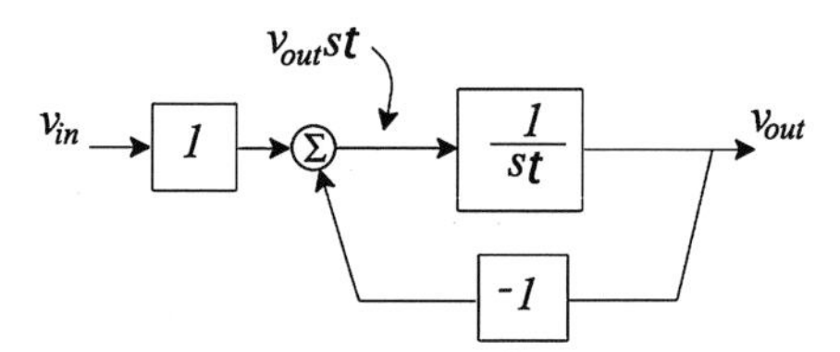

Fig. 6.57 Flow graph for the low-pass filter generated from the transfer function in Eq. (6.47) starting from the component of v_{out} with the highest power of "s"

using the Z-method form for feedback analysis. The low-pass transfer function is solved for the v_{out} term with the highest power of "s," Eq. (6.47).

$$H_{\text{lp}}(s) = \frac{1}{1 + s\tau} = \frac{v_{\text{out}}}{v_{\text{in}}}$$
$$v_{\text{out}} \cdot s\tau = v_{\text{in}} - v_{\text{out}} \tag{6.47}$$

From the final line in Eq. (6.47), we generate the flow graph as shown in Fig. 6.57 that can be interpreted as the realization in Fig. 6.56. This suggests a design process, a mapping from the transfer relation to, first, a flow graph in final transfer function parameters; second, to a circuit topology using G_m–C elements; and third, solving for parameters in the real design by equating coefficients in both transfer functions. From Eqs. (6.46) and (6.47), we see that we make $\tau = c_1/G_m$ to realize the low-pass filter having a cutoff frequency of $1/(\tau \times 2\pi)$.

The second-order transfer function is a bit more challenging. We can map the transfer function in Eq. (6.48) to the flow graph in Fig. 6.58:

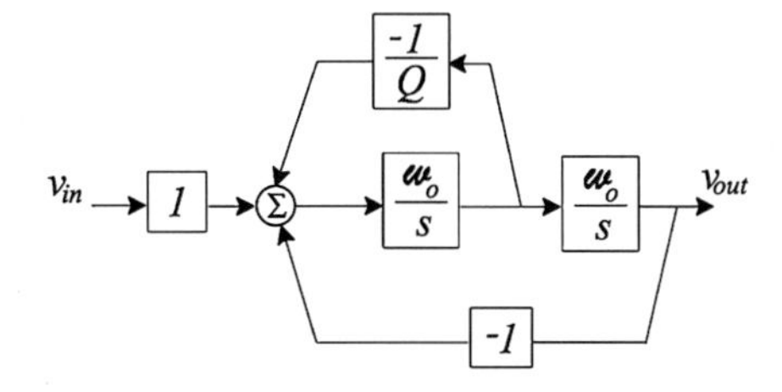

Fig. 6.58 Flow graph for the second-order filter generated from the transfer function in Eq. (6.48) starting from the component of v_{out} with the highest power of "s"

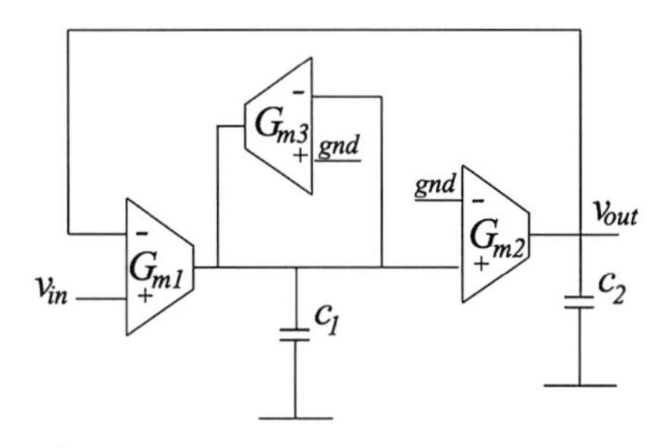

Fig. 6.59 G_m–C realization of second-order transfer function in Eq. (6.48)

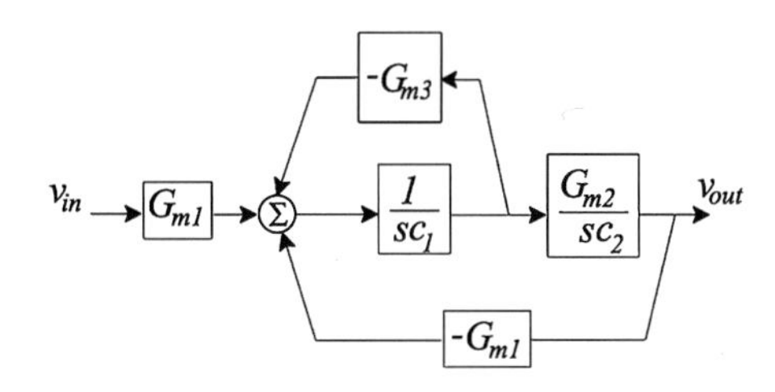

Fig. 6.60 Flow graph for G_m–C realization of second-order transfer function in Fig. 6.59

$$H_{2^{\mathrm{nd}}\,\mathrm{Ord}}(s) = \frac{1}{1 + \frac{s}{\omega_0 Q} + \frac{s^2}{\omega_0^2}} = \frac{v_{\mathrm{out}}}{v_{\mathrm{in}}}$$

$$v_{\mathrm{out}} \cdot \frac{s^2}{\omega_0^2} = v_{\mathrm{in}} - v_{\mathrm{out}}\left(1 + \frac{s}{\omega_0 Q}\right) \tag{6.48}$$

Once we have the flow graph for the transfer function we want to realize, we interpret it as the G_m–C filter shown in Fig. 6.59. This schematic then gets mapped as the flow graph in Fig. 6.60.

Once we have the flow graph for the real circuit, we find its transfer function and equate corresponding terms to find the particular design parameters, Eq. (6.49):

$$H_{2^{\mathrm{nd}}\,\mathrm{Ord}}(s) = \frac{1}{1 + \dfrac{s}{\omega_0 Q} + \dfrac{s^2}{\omega_0^2}}$$

$$v_{\mathrm{out}} = \frac{1}{1 + \dfrac{sc_2 G_{\mathrm{m3}}}{G_{\mathrm{m1}} G_{\mathrm{m2}}} + \dfrac{s^2 c_1 c_2}{G_{\mathrm{m1}} G_{\mathrm{m2}}}} \tag{6.49}$$

$$\frac{1}{\omega_0^2} = \frac{c_1 c_2}{G_{\mathrm{m1}} G_{\mathrm{m2}}};\ \frac{1}{\omega_0 Q} = \frac{c_2 G_{\mathrm{m3}}}{G_{\mathrm{m1}} G_{\mathrm{m2}}}$$

We see in Eq. (6.49) that we have two equations but five unknowns. We can make the three G_m's equal to each other and, from a power specification, design an OTA to have that transconductance, leaving two unknowns that we can now solve.

With these building blocks, and integrator, the first-order, and the second-order section, we can build multiple pole transfer functions. A transfer relation may contain zeros. Adding a zero to the single-pole transfer function and manipulating it to show v_{out} with the highest power of "s," we get Eq. (6.50) which we map as the flow graph in Fig. 6.61 and then onto the G_m–C realization in Fig. 6.62.

$$H_{\mathrm{pz}}(s) = \frac{1 + s\tau_{\mathrm{z}}}{1 + s\tau_{\mathrm{p}}} = \frac{v_{\mathrm{out}}}{v_{\mathrm{in}}}$$
$$v_{\mathrm{out}} \cdot s\tau_{\mathrm{p}} = v_{\mathrm{in}}(1 + s\tau_{\mathrm{z}}) - v_{\mathrm{out}} \tag{6.50}$$

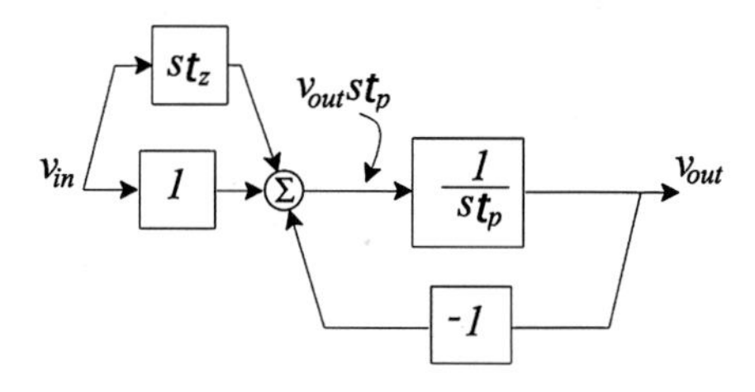

Fig. 6.61 Flow graph for a transfer function with a single pole and a single zero

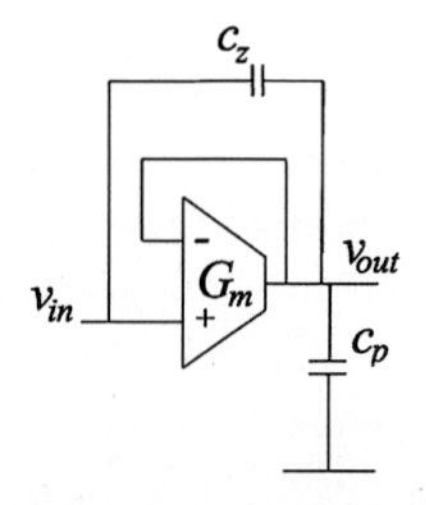

Fig. 6.62 G_m–C realization of a transfer function having a zero and a pole

The zero is added through the feedforward capacitor. This G_m–C schematic has the transfer function given in Eq. (6.51):

$$H_{\text{pz}}(s) = \frac{1 + s\tau_{\text{z}}}{1 + s\tau_{\text{p}}} = \frac{v_{\text{out}}}{v_{\text{in}}}$$

$$v_{\text{out}} = \frac{(G_{\text{m}} + sc_{\text{z}})}{G_{\text{m}} + sc_{\text{z}} + sc_{\text{p}}} v_{\text{in}} = \frac{1 + s\dfrac{c_{\text{z}}}{G_{\text{m}}}}{1 + s\dfrac{c_{\text{z}} + sc_{\text{p}}}{G_{\text{m}}}} v_{\text{in}} \tag{6.51}$$

$$\tau_{\text{z}} = \frac{c_{\text{z}}}{G_{\text{m}}};\ \tau_{\text{p}} = \frac{c_{\text{z}} + sc_{\text{p}}}{G_{\text{m}}}$$

Equating corresponding factors, we solve for the two capacitors in terms of G_m that we design to meet performance requirements. A general filter transfer function can now be constructed using one and two pole blocks plus feedforward capacitive paths in the flow graphs and then in the G_m–C realizations.

Summary

This chapter has used the DPI/SFG approach coupled with the Z-method for loop gain to analyze various types of circuits in order to demonstrate the versatility of this methodology. The circuits were mostly analyzed by inspection, mapping the specific topology onto a flow graph, and then doing graph manipulations to obtain the desired transfer function. The process breaks down the original circuit into more manageable pieces that can then be interpreted as a signal flow graph for analysis. Basic building blocks were examined first—current mirrors, reference cells, and an OTA. Other cells followed using the same process: voltage and current references, a capacitive gain cell, a low dropout regulator followed by a switching regulator, phase-locked loop design, and finally G_m–C filters. All of these examples have feedback and were handled directly using the Z-method. Analysis of feedback systems has been shown to be handled using a direct extension of the DPI/SFG method, the Z-method, with particular attention given to the signal path—whether it flows in the loop or flows to ground. Once this separation is done, feedback is handled unambiguously and consistently. Further, application of this approach leads to the development of insight into circuit design and analysis with the focus on the topology leading to the transfer relations in factored form providing insight to design. Poles and zeros are associated with particular nodes in the design and can be individually addressed. The use of hand analysis for zero-order analysis becomes manageable and orderly. CAD tools are then used for examination of higher-order behavior and for verification. Systems using transducers or changes in variables are also addressed using this approach. Phase-locked loops and switching regulators use signal averages and changes in signal types to process the top-level system behavior and are consistent with flow graph analysis.

References

1. P.R. Gray, R. Meyer, *Analysis and Design of Analog Integrated Circuits* (Wiley, New York, NY, 1977)
2. Behzad Razavi, *Design of Analog CMOS Integrated Circuits* (McGraw Hill Publishing, New York, NY, 2000)

Index

A. Ochoa, *Feedback in Analog Circuits*, DOI 10.1007/978-3-319-26252-9

Printed by Printforce, the Netherlands